"Bruges' book is up-to-date and as comprehensive as any book could be at this extremely early stage of interest in the issue."
—DENNIS MEADOWS, coauthor of *Limits to Growth* and Emeritus Professor of Systems Policy and Social Science Research, University of New Hampshire

"It's not enough to stop burning fossil fuels. We also have to remove much of the carbon dioxide that has accumulated in the atmosphere for over a century. Biochar is one of the few tools available for that purpose. If you don't know what biochar is, this book tells you what you need to know."
—PETER BARNES, author of *Climate Solutions* and *Capitalism 3.0*

"*The Biochar Debate* is an intelligent and evenhanded look at the potential for both improving soil and addressing global warming offered by the decentralized production and use of biochar. The potential pitfalls and unknowns are clearly acknowledged—this is not another faddish silver-bullet approach, but offers some real-world examples and practical ideas that anyone can use."
—GRACE GERSHUNY, coauthor of *The Soul of Soil*

"The buzz of interest and activity around biochar in recent years is accelerating. In this concise but engaging book, James Bruges gets us up to speed with the ecology, economics and politics of biochar. Over three decades of speaking about and teaching permaculture, I have come across very few sustainable 'technologies' that appear to change the rules about how to work with nature. Biochar is one of those few. Could biochar be the simple solution by which we can save civilization from the twin crises of resource depletion and climate catastrophe? This sounds like an absurd claim, but not one that can be easily dismissed. James Bruges steers a course between the hope and the hype."
—DAVID HOLMGREN, co-originator of the Permaculture concept and author of *Future Scenarios*

"Our planet is in an existential crisis. While scientists fret and economists debate, politicians dither and business leaders derail. There is a disconnect between physical reality and political reality. And yet, the physical one always trumps; did we imagine it otherwise? James Bruges has got this right. Biochar offers us a last chance to cheat death, but we'll be given only one try. Fail and our epitaph will be a hard, black layer writ in the strata: Here Lies the Human Experiment, R.I.P."

—ALBERT BATES, author of *The Post-Petroleum Survival Guide and Cookbook* and founder of Global Village Institute for Appropriate Technology

"A brilliant, readable review on the critical need to restore our degraded lands back to fertility—be it to sequester greenhouse gases naturally, support forests, improve soil moisture, or increase crop yields. Bruges outlines how supporting *natural* terrestrial sequestration is the cost-effective, proven practice to extract carbon from the atmosphere, and that this can be augmented via the use of soil amendments such as biochar. He concludes with examples that elucidate why tying biochar-based land-management solutions to one-size-fits-all market incentives risks time, money, and public health. Our students say, 'It's a 101 must-read'—a strong recommendation, indeed."

—ALISON BURCHELL, geologist, Natural Terrestrial Solutions Group

"Biochar is a relatively new word in the green lexicon, but one you'll hear more about going forward. It isn't a silver bullet, but it may be a useful help in the climate challenge—this slim book will let you think knowledgeably about it, and start to act in your own backyard."

—BILL MCKIBBEN, author of *Eaarth: Making a Life on a Tough New Planet*

"A brilliant synthesis for everyone concerned with solutions to climate change, enhancement of our soils, and the future of energy policy. An enjoyably readable introduction to the vital field of biochar. Highly recommended."

—L. HUNTER LOVINS, founder and president of Natural Capitalism Solutions, cofounder of the Rocky Mountain Institute, and coauthor of *Natural Capitalism*

Schumacher Briefing No. 16

the biochar debate

charcoal's potential to reverse climate change and build soil fertility

James Bruges

with biochar research and
development in southern India
and illustrations
by
David Friese-Greene

Chelsea Green Publishing
White River Junction, Vermont

Scanning Electron Microscope (SEM) photographs © Dr. Stuart Kearns of the Earth Sciences Department of Bristol University.
All other illustrations © David Friese-Greene.

First published in 2009 by Green Books Ltd
Foxhole, Dartington, Totnes, Devon TQ9 6EB
United Kingdom
www.greenbooks.co.uk

for The Schumacher Society
The CREATE Centre, Smeaton Road,
Bristol BS1 6XN United Kingdom
www.schumacher.org.uk •
admin@schumacher.org.uk

The Schumacher Briefings
Series Editor: Stephen Powell
Founding Editor: Herbert Girardet

Printed in the United States of America
First Chelsea Green printing January, 2010
10 9 8 7 6 5 4 3 2 1   10 11 12 13 14 15

Our Commitment to Green Publishing
Chelsea Green sees publishing as a tool for cultural change and ecological stewardship. We strive to align our book manufacturing practices with our editorial mission and to reduce the impact of our business enterprise in the environment. We print our books and catalogs on chlorine-free recycled paper, using vegetable-based inks whenever possible. This book may cost slightly more because we use recycled paper, and we hope you'll agree that it's worth it. Chelsea Green is a member of the Green Press Initiative (www.greenpressinitiative.org), a nonprofit coalition of publishers, manufacturers, and authors working to protect the world's endangered forests and conserve natural resources. *The Biochar Debate* was printed on American Eagle White, a 30-percent postconsumer recycled paper supplied by Thomson-Shore.

Library of Congress Cataloging-in-Publication Data
Bruges, James.
   The biochar debate : charcoal's potential to reverse climate change and build soil fertility / James Bruges, with biochar research and development in southern India and illustrations by David Friese-Greene.
      p. cm. -- (Schumacher briefing ; no. 16)
   Includes bibliographical references and index.
   ISBN 978-1-60358-255-1
   1. Ashes as fertilizer. 2. Charcoal. 3. Soil amendments. 4. Carbon sequestration. 5. Climatic changes--Prevention. I. Title. II. Series.

S663.B78 2009
631.8--dc22

                            2009046545

Chelsea Green Publishing Company
Post Office Box 428
White River Junction, VT 05001
(802) 295-6300
www.chelseagreen.com

Contents

Chapter 1

Introduction

Charcoal and biochar

Charcoal is one of the oldest industrial technologies, perhaps the oldest. In the last decade there has been a growing wave of excitement engulfing it. Why?

Because some scientists are saying that we might be saved from the worst effects of global warming if we bury large quantities of it. Not only that: we can restore degraded land and get better harvests by mixing fine-grained charcoal—biochar—with soil. Others say that charcoal's use could be just one of several technologies to mitigate climate change. Yet some maintain that it is an extremely dangerous technology. The jury is out on which is closest to reality. This Briefing aims to provide an overall view of the subject and describes the best way to encourage the appropriate use of biochar.

The theory is simple. Plants, through photosynthesis, capture carbon dioxide—the main greenhouse gas—from the air as they grow. The carbon of CO_2 provides their structure and the oxygen is released for animals to breathe. If the plants are left to rot, the C and O combine again in a relatively short time to release carbon dioxide back into the air. However, if the plants are heated in the absence of oxygen—called pyrolysis—charcoal is formed. Charcoal is largely carbon. As anyone who has organized a barbecue knows, charcoal can be burned, in which case the carbon goes back up into the atmosphere. But if it is buried, the two elements take a long time to recombine as carbon dioxide. This means that some of the most abundant greenhouse gas can be taken out of the atmosphere and locked into the ground for a long time. Deep burial—rather like putting coal back where it belongs—is one way. But there is another option.

Additional excitement came with the discovery of deep dark areas of *"terra preta do indio"*—Indian black earth—in the Amazon rainforest where the soil generally is thin, red, acidic and infertile. The patches of *terra preta* are alkaline with a high carbon content, and contain pot-shards indicating that it was not natural: a pre-Columbian civilization had created it. It is extracted and widely used by garden contractors because it is so fertile. It has remained fertile and retained its carbon content through the centuries.

Terra preta is black because it contains large amounts of charcoal. Infertile land had been converted to fertile land that supported a thriving civilization through the wise use of the trees that had been felled. Could charcoal, therefore, not only be a vehicle for reducing global warming but also a means to increase the fertility of degraded land, and help feed the world?

Charcoal used for this purpose is referred to as biochar. Biochar is pulverised charcoal made from any organic material (not just wood) and, when mixed with soil, it enhances its fertility. It locks carbon into the soil and increases the yield of crops. To many, this appears the closest thing to a miracle.

The process of converting plant material to charcoal gives off heat together with gases and oils. Certain plants and certain processes produce a high proportion of charcoal, whereas others produce more gases and oils. This is where the problems start. These chemicals could become the main commercial attraction of biochar. As has been found with biofuel, growing crops to fuel cars can be more profitable than growing food to feed people. If left to the market, producers of biochar might buy up productive land, plant monocultures, and develop their equipment primarily to produce fuel and industrial chemicals.

Then there is the suggestion that the burial of charcoal should earn carbon credits. As above, the financial motive could lead to "growing carbon credits" in preference to growing food. And if widely adopted, as hoped, the carbon market would be flooded with credits; industry would buy them at fire-sale prices and carry on with business as usual to the detriment of the climate. A strong financial incentive to use biochar is desirable, but carbon credits may not be the best approach.

There are two prime objectives. It is essential to find ways to sequester greenhouse gases if we are to avoid the worst effects of global warming. It is essential also to enable farmers throughout the world to use biochar if it can bring degraded land back to fertility and increase yields. The process cannot be left to "the market," which has been described as an out-of-control demolition ball swinging from a high crane.

In the final chapter I outline twin policies for reducing greenhouse gases in the atmosphere. The first policy would ensure a reduction in the use of fossil fuels. The second concerns our use of land. The

A typical Indian village.

requirements for the two are so different that separate regulations are necessary. The first is called cap-and-dividend (in the US). The second is the Irish proposal for a Carbon Maintenance Fee (CMF), which would provide a powerful incentive for every country, rich and poor, to enable its farmers, businesses and individuals to maintain land-based carbon.

A visit to India

I had been caught up in the excitement surrounding biochar before visiting India in January 2008. While there I asked everyone I met about the production and use of charcoal. My obsession was embarrassing for my wife Marion, but paid off when it led to meeting Dr. Ravikumar of the Centre for Appropriate Rural Technology (CART) in Mysore. He talked about a "charcoal revolution" that would bring employment to the rural poor. He had been work-

Demonstrating Ravikumar's Anila stove.

ing on stoves to produce charcoal in the absence of oxygen for about eight years at CART. But CART was closing, so he was looking for an organization to promote the development of his ideas. I suggested he contact our friends Amali and Cletus Babu, who had started the non-governmental organization SCAD, Social Change and Development, in the southern tip of India 25 years earlier. Their objective is also to bring knowledge and employment to communities in the 450 villages where SCAD works.

Later, David Friese-Greene, while visiting SCAD, was taken to meet some banana growers who had been adding charcoal dust to their crops during the last four years. They had tried this quite by chance, having come across a supply of almost free charcoal made from rice-husks as a by-product of some other process. They told David that digging it in with the banana plants cut the amount of irrigation water needed in half and doubled the yield of their crop. Maybe there was a bit of exaggeration here—I don't know. They added that the bananas taste better. In Britain we only receive one kind of banana, referred to by our Indian friends as tasting like expanded polystyrene, but there is a great variety in south India and the different tastes are appreciated and affect sales. Neighbouring farmers have been impressed with the results and are adopting their practice.

The fact that burying charcoal extracts carbon dioxide from the atmosphere is of little significance to SCAD's farmers. They are only interested in increasing the yield of their crops. If this can be clearly demonstrated to, or by, farmers, and if the equipment for producing the charcoal were affordable and available, the practice would spread naturally.

But everyone in India is aware of carbon credits. They are like a magic wand, a source of income and development whether or not they are truly effective in reducing carbon emissions, or whether they simply allow Western industry to carry on emitting as usual. So we discussed the subject.

If the farmers were to earn carbon credits, surely this would be an added benefit for them? Possibly; but possibly not. There is no way in which these farmers could take part in transnational economic mechanisms. As the value of credits rose to exceed the value of food crops, the smallholders would be displaced by agribusiness seeking to accumulate large areas and plant monoculture crops primarily for the purpose of attracting credits. Only agribusiness could handle the complicated global trade. The process would require an army of monitors, but this would be an open invitation for corruption. Western entrepreneurs looking to extract profit would descend like hawks. And, of course, setting up this novel kind of infrastructure would take years. In a small way in this corner of India I had become aware of the arguments taking place more widely between advocates and opponents of biochar.

The players

The Rio Earth Summit in 1992 was, quite possibly, the most remarkable meeting of leaders that has ever taken place. Heads of state met specifically to respond to concerns raised by scientists about global warming. For the previous seven years scientists had been saying unequivocally that this was due to emissions caused by human activity and urgent action must be taken. The remarkable thing is that these leaders listened to scientists. Unfortunately, politics subservient to reality did not continue at subsequent climate meetings, where negotiators seemed to do all they could to obscure the science.

At Rio, the world leaders set an agenda "to achieve . . . stabilization of greenhouse gas concentrations in the atmosphere at a level that would prevent dangerous anthropogenic interference with the climate system." Direct action by governments during the following years could have saved us from the present crisis.

However, market fundamentalism intervened. The US and Britain were intent on subjecting all aspects of the global economy to growth, free trade, and neoliberal economics. Democracy was able to assert minimal restraint over the power of corporations and lobby groups. The result has been a disaster for both the economy and the climate. It was in this mental environment that the follow-up meeting took place at Kyoto in 1997. Negotiators listened to economists and industry, not to scientists. Action would be taken by "market mechanisms," not by strict rules. The regulation of emissions would become a global trade with new opportunities for profit.

The result? The emission of greenhouse gases, far from reducing, has been increasing at an accelerating rate, and it is no longer possible to achieve the full requirements of Rio; we can only hope to avoid the worst effects. Can this be described as anything other than failure? There are plenty of excuses of course. "It is the only game in town," they now say. But nature doesn't listen to excuses. Failure, eighteen years since Rio. Failure, thirteen years since Kyoto.

The uncertainties associated with biochar—proof of its permanence, its effect on forests and crops, the million different ways in which it might be used, certification, the money trail—are more complex than negotiators have ever previously encountered. Coming to a global agreement on carbon credits for charcoal would take years. And if it were successful, the carbon market would be flooded with credits and collapse. Industry would then have little restraint on its emissions. How many more years of failure before the rich are prevented from buying indulgences that allow them to carry on making yet more money while destroying the climate? And how much time do we have? The legacy of national leaders will not be determined by their handling of war, terrorism, or the economy but by their failure to take immediate action over the threat of global warming. They could initiate measures to encourage biochar within a month, given the will.

The aforementioned governments and global negotiators are one group of players. In addition there are many groups involved in promoting the use of charcoal for sequestration that are primarily concerned with understanding the science, demonstrating the benefits, and promoting good practice. The International Biochar Initiative (IBI) is a leading body promoting scientific and practical understanding.

The most thorough scientific study so far is *Biochar for Environmental Management*, edited by Johannes Lehmann and Stephen Joseph, and I refer to it frequently. Chapters are written by scientists for scientists— each concentrating on a single aspect—demonstrating the ecological complexity of biological systems and, incidentally, vividly showing how difficult it would be to define, for commercial purposes, its effectiveness in sequestering carbon dioxide. The book has few direct hints for farmers and is rather impenetrable for the general reader.

There is a need for an intermediate organisation to translate theory into advice, in the way that the USDA-funded National Sustainable Agriculture Information Service advises organic farmers. Practitioners using trial-and-error methods throw up complex, messy, and difficult questions that cannot be resolved simply by reference to the science.

Then there are commentators who talk of economics and regulation. However, scientists, practitioners, and commentators all agree that a great deal of research and trial will be necessary for a full understanding and application of the process. Scientific uncertainty over the application of biochar, however, is no reason to rubbish the whole subject. By doing this the biochar skeptics lose credibility for their reasoned campaign against the market-based approach. Remember that uncertainty over detail was used to great effect by climate skeptics who wished to rubbish the science of climate change.

Biofuelwatch is an organisation that has highlighted the dangers of biofuels and argues that many of the same dangers apply to the development of biochar. Some industrialists see charcoal just as a by-product of a process that is primarily focused on producing biofuel. Companies developing biofuel aim to replace fossil fuels with "clean" energy from plants, though the claim to be clean has proved faulty in many cases. It is also generally accepted that the rise in food prices above what many of the poor can afford is due largely to biofuels displacing grain crops. Biofuelwatch examines what would happen if the production and burial of charcoal were to earn carbon credits under the Kyoto Protocol or its successor. As the availability of fossil fuels diminishes, the value of these credits could become so huge that commercial interests would initiate planetary geo-engineering projects—Shell and J. P. Morgan are already in on the act—and farming could give way to industrial monocultures that would have unknown but potentially disastrous consequences for the climate, for people, and for biodiversity.

Then there is James Lovelock. In a recent interview in *New Scientist* replying to the question "Are we doomed?" he said, "There is one way we could save ourselves and that is through the massive burial of charcoal. It would mean farmers turning all their agricultural waste—which contains carbon that the plants have spent the summer sequestering—into non-biodegradable charcoal, and burying it in the soil. Then you can start shifting really hefty quantities of carbon out of the system and pull the carbon dioxide down quite fast. . . . This scheme would need no subsidy: the farmer would make a profit." Note that his focus is on farming, not on economic mechanisms.

Chris Goodall includes biochar in his excellent book *Ten Technologies to Save the Planet*. This analyses each of the technologies from a commercial perspective. Only two of the technologies would extract carbon dioxide; the others would just reduce emissions. There are many other books and magazine articles that touch on the subject. In his book *Sustainable Energy—Without the Hot Air* Professor David MacKay analyses the *potential* sustainable fuels for Britain. He describes his approach as "numbers not adjectives"—i.e. setting aside preconceptions, economics, politics, and even ethics. He analyzes the potential and limitations of each permanent energy source, with solar photovoltaics emerging as having the greatest potential. It is interesting that 70 percent of renewable energy will reach the end user as electricity. This is important for determining the type of equipment that should be researched and developed. One of the biggest savings as compared with present energy use will be in doing away with the need to convert fossil energy into electricity.

In all of this I have been influenced by the philosophy of Fritz Schumacher, best known for his classic work, *Small is Beautiful*. If his recommendations had been followed, we would not be in the mess we are in today. I have found much in his work that is directly relevant to the use of biochar, so I have tried to look at the issues through his eyes.

As you can see from the above, this Briefing is not just about the stuff we burn in barbecues. A very old technology—the production of charcoal—in the hands of farmers throughout the world could become a major player in the struggle to avoid the worst effects of global warming. Given such potential, it must be considered in relation to the global economy, to commercial pressures, to international negotiations, and not least in relation to agricultural practice.

Chapter 2

An Overall View

"There is one way we could save ourselves [from global warming] and that is through the massive burial of charcoal."

–James Lovelock, New Scientist, 24 January 2009

Carbon

Charcoal is largely carbon. Carbon is essential to life on earth. In fact, carbon constitutes the very definition of life because its presence or absence indicates whether a molecule is organic or inorganic. Every organism needs carbon either for structure, energy or, as in the case of humans, for both. Discounting water, about half your body weight is carbon.

Carbon is the fourth most abundant element in the universe. Water vapor and carbon, in the form of carbon dioxide, are the two most pervasive greenhouse gases that blanket the earth and prevent it becoming too cold for life.

Carbon in fossil fuels has been our main source of energy. By burning fossil fuels and releasing their carbon we are causing global warming and, since these are limited resources, we are coming to the end of the cheap energy on which modern prosperity is based. Concern over climate and energy are linked by carbon.

If charcoal can play a major part in avoiding catastrophic global warming, discussion of its use must take place in the context of the atmosphere, the land, fossil fuels, and much else.

The atmosphere

"We have at most ten years to make drastic cuts in emissions that might head off climate convulsions."

—James Hansen, NASA Goddard Institute for Space Studies, 2005

There have been five mass extinctions since animals first evolved. A meteorite caused the one that did away with the dinosaurs 65 million years ago. The previous four are referred to as "microbial." The Permian extinction was the result of excessive carbon dioxide from volcanoes blanketing the planet, which led to runaway global warming. The poles warmed faster than the rest of the planet. This reduced temperature differences between latitudes and caused ocean currents to stall. Without these circulations the depths were deprived of oxygen. Rotting life forms in the stagnant oceans belched out hydrogen sulphide, which drifted over the land killing most terrestrial life. Without roots to hold it, soil washed into the seas. The Earth took millions of years to recover. Changes to the climate that are happening now have similarities with some aspects of these microbial mass extinctions.

Our planet would be without life if it were not for the delicate carbon cycle between plants and atmosphere. James Lovelock refers to the living planet as Gaia, though he is sometimes embarrassed that this metaphor from the Greek goddess of the Earth—who was the most revered goddess of all and the only one who never misbehaved—detracts from the strict science of Earth as a self-regulating system. Plants interact with minerals, air, and water in ways that we are only beginning to understand, and their ability to extract carbon dioxide from the air is a key element in regulating the climate. They keep the surface of our planet at a temperature that is not cold like the moon—our "natural" temperature—and not too hot. The five mass extinctions, together with ten other minor ones, demonstrate that this delicate system is fragile. If disturbed excessively, it can tip into catastrophe.

Every 100,000 years in recent geological history there has been an interglacial period, kick-started when our planet orbited closer to the sun, and maintained by an increase in the carbon cycle as plants replaced ice. The present interglacial period started about 10,000 years ago and is referred to by geologists as the Holocene epoch, which enabled humans to settle and cultivate. We are due for another glacial period, but the reverse is likely to happen. Human activity destroyed many of the forests and, in recent years, released excessive quantities of carbon that had been safely fixed in geological strata. This new development in the Earth's history is being referred to as the Anthropocene epoch.

Over the last 250 years, particularly the last 50, we have been mining carbon that had been extracted from the atmosphere by plants millions of years ago and buried as coal, gas and oil. By burning it, the carbon combines with oxygen to form carbon dioxide. It is predicted that the consequent warming will cause floods, droughts and mass migration. But more alarming are parallels with the Permian extinction. The poles are warming faster than elsewhere. The Gulf Stream is slowing. And, due to feedback effects, a tipping point might be reached fairly soon when runaway global warming will take over and we will be unable to stop it. Many scientists say that global warming is the greatest threat humanity has ever faced.

The complacency of politicians is difficult to comprehend. I live in the UK where annual expenditure on climate issues is measured in a few millions; on "defense"—while admitting we have no external threats—in billions (£36bn [$57bn]); and on banks? Politicians have signed a check for £1 and allowed the recipients to add zeros. In the US the disparity is even more extreme. The situation is so dire James Hansen has said, "Civil resistance is not an easy path, but given abdication of responsibility by the government it is an essential path."

If the process of turning plants into charcoal can extract carbon dioxide from the atmosphere on a large scale, it should be right at the top of priorities for scientists, politicians, economists and, of course, farmers.

The land

Soil, originally rock, has taken thousands of years to form. When considering the period of settled communities, Schumacher quoted a passage in *Small is Beautiful* from the book *Topsoil and Civilization* by Tom Dale and Vernon Gill Carter: "Civilised man has despoiled most of the lands on which he has lived for long. This is the main reason why his progressive civilisations have moved from place to place." He adds, "The 'ecological problem', it seems, is not as new as it is frequently made out to be. Yet there are two decisive differences: the earth is now much more densely populated than it was in earlier times and there are, generally speaking, no new lands to move to."

In this passage Schumacher identified two big issues of our time. First, it is questionable whether the planet can support the huge increase in population that has occurred over the last 100 years. So, if we are to avoid a Malthusian nightmare, all countries will need to adopt the specific conditions that are causing fertility rates to decline in some countries. The low fertility rate of Kerala, one of the least developed of India's states, shows that a stable or reducing population can result more from women's education than from financial prosperity: Kerala has a literacy rate approaching 100 percent. Second, since humans have degraded the land for millennia, and particularly during the last half-century, we are facing the fate of many previous civilizations. They thought that politics, social organization, technology, economic acuity or military power could sustain them, just as we do. Schumacher suggests that our efforts should, instead, turn to helping the living planet, Gaia, restore those parts of the land that have been degraded.

One of the greatest benefits of biochar is its ability to transform degraded land. It adds moisture retention to otherwise near-desert conditions, it provides surfaces for microbes and nutrients to use, and it can lock carbon into the ground. It has special value for tropical conditions, which implies that smallholder farmers in poor countries could benefit most, but it could also be critical in restoring soil in temperate climates that has been degraded by synthetic chemicals.

Limited resources

"The unexpected legacy of fossil fuels leads us to lose sight of the principle of a durable economy, which needs to be based exclusively on the regular influx of energy from the sun's radiation."

—Wilhelm Ostwald, Nobel Prize for Chemistry, 1909

Western society's big mistake was basing its entire economy, and the global economy, on a limited resource.

This seems obvious now that the decline of this resource approaches. But is it only obvious with hindsight? I have found it fascinating to look for the views of reflective thinkers at the time when fossil fuels started to drive the economy. The result? It has indeed been obvious to scientists for over a century. If statesmen had listened to scientists, we might have taken a different path. We are learning to our cost, however, that politicians like to play with military power and consult commerce and neoliberal economists in preference to listening to scientists—even now. This is a worrying conclusion when scientists are warning of cataclysmic dangers.

It used to be thought that transparent gases were also transparent to heat. In 1859, the year Brunel died, the Irish scientist John Tyndall was in London suffering the grime of smoke from thousands of coal fires and steam trains. In front of a packed audience at the Royal Institution, with Prince Albert in the chair, he demonstrated that carbon dioxide absorbs heat over a wide spectrum though oxygen and nitrogen do not. He concluded: "The radiant heat of the sun does certainly pass through the atmosphere to the earth with greater facility than the radiant heat of the earth can escape into space." This linked the burning of fossil fuel with the climate.

In 1896 Svante Arrhenius, who was to become a founding director of the Nobel Institute for Chemistry, coined the phrase "greenhouse effect" and predicted that if concentrations of carbon dioxide in the atmosphere doubled, the global climate would warm by $4°–6°C$ ($7.2°–10.8°F$). These figures are remarkably close to recent analysis. A bit later, in 1908, he saw clearly why the wrong path would be chosen. "Every industrialist seeks to push his production as high as possible . . . and he gives no thought whatsoever to how things will be in half a century. The statesman, however, needs to apply a different standard."

His warning applies equally today for the development of biochar if it is to be a major technology for sequestering carbon dioxide. If left to market mechanisms, commercial development will not be driven by the need to extract carbon dioxide from the air but by the need to maximize profit. To start with we may fool ourselves that the two are synonymous, but this approach could lead to perverse unintended outcomes.

Right at the beginning of the exploitation of oil, Wilhelm Ostwald was even more specific. He pointed to the madness of basing our entire economy on an energy resource that could not be maintained. His warning gave us a full century to reorient our economy around regular energy from the sun.

I came across further evidence that scientists consistently held this view quite by chance. Jeremy Cullen, a friend, had been given a library assembled by Sir Edgar Sylvester, first Chairman of the Gas Council. Among the books was a rather scruffy little orange-striped Penguin, obviously well used, called *Science and World Order*. Jeremy happened to notice it and thought I might be interested. I was!

It records the proceedings of a conference in 1941 at the Royal Institution, attended by scientists from twenty countries including the United States, Soviet Union, China, India, South Africa and many European countries, and also Sir Anthony Eden and H. G. Wells. The purpose was to decide what policies needed to be adopted after the war.

Being in central London during the Blitz must have concentrated their minds. One chapter is on world resources, particularly fossil fuels, and this is one of the comments contrasting finite and renewable resources: "Uncontrolled exploitation of resources may hamper our descendants. Certain resources, once consumed can never be replaced, whilst others are renewable. In making use of the first kind we are living on capital, but in using the second we are only consuming income." Natural gas and oil were recognized as having a limited future. The chapter then contrasts non-renewable—coal, gas, and oil—and renewable energy with a discussion of hydropower, tides (including the Severn Barrage), geothermal, biofuel, solar photoelectric, solar photochemical, solar steam generation using mirrors, and talks about North African deserts as centers of power distribution. The fact that most of these renewable energy sources are still being talked about and are still in their infancy 70 years later indicates a huge lost opportunity. If politicians had listened to these scientists, they would have chosen a different path for post-war reconstruction.

But let's go on. Half way through the "oil era," in 1956 (over half a century ago), the chief scientist of Shell, Marion King Hubbard, warned that the extraction of oil in the United States would peak within fifteen years. As the extraction of oil increased during those fifteen years, people laughed at him. The peak was not obvious at the time because the graph of extraction had flattened to be a sort of plateau. Only in retrospect was it clear that his prediction had been right. He subsequently said that world oil would peak at the turn of the century. It probably would have were it not for the politically induced slow-down in the 1970s. It reached a plateau in 2005 at about 74.8mb/d, and the amount extracted has not increased in spite of rising demand and the most sophisticated equipment. The International Energy Agency and the respected investment firm Raymond James suggest that the "peak" will be seen to have taken place in early 2008 and the decline will start in 2012.

One more comment, this time from the political side. UK Prime Minister Gordon Brown, speaking in 2008, said: "It is a scandal that OPEC can restrict the supply of oil at a time when oil is desperately needed." Isn't the scandal that, after 100 years' warning, only 2 percent of UK energy is derived from permanent sources? It's crazy to continue with our economy and agriculture totally dependent on oil. Producers will increasingly be able to "hold us over a barrel." The longer the change is delayed, the steeper and more catastrophic will be the decline. Our need to extract carbon dioxide from the atmosphere—with charcoal as the main facilitator—is a result of basing our entire economy on the limited resource of oil.

Geo-engineering

"Do we really believe that we humans, untrained as we are, have the intelligence or capacity to manage the Earth?"

—James Lovelock, 2009

In 1972 Edward Lorenz gave a talk entitled, "Does the flap of a butterfly's wings in Brazil set off a tornado in Texas?" The phrase, which became famous, developed from an incident that happened ten years earlier. To save time, he had entered a simplified number into a computer model of the weather—0.506 instead of 0.506127, a tiny difference—and the computer threw up a completely different weather scenario. You can't say that the butterfly caused the tornado but, under chaos theory, the flapping wing made a small change to the initial condition of a system that caused a chain of events leading to a large-scale alteration of outcomes. Similarly, climate-skeptics have difficulty in imagining that adding two parts per million of carbon dioxide to the atmosphere annually can be the cause of catastrophic global warming.

"Engineering" means a deliberate attempt to apply scientific principles to achieve a project. In the context of the world's climate, the intention may be one thing and the outcome another. The intervention may be small but the effect unpredictably large. This suggests that extreme caution is necessary when interfering with natural climate systems.

Agriculture can be regarded as one such intervention. Humanity turned to agriculture a few millennia ago, a fraction of the time-span of our species. The objective was food, but to achieve this forests were destroyed and biodiversity harmed. "The problem is that man's conquest of the world has itself devastated the world. And in spite of all the mastery they've obtained, they don't have enough mastery to stop devastating the world—or to repair the devastation they've wrought." This was the gorilla, Ishmael, speaking in Daniel Quinn's wonderful 1992 book on humanity's adoption of agriculture. His pupil, having finally accepted the summary, replied: "And given a story to enact in which the world is a foe to be conquered, we will conquer it like a foe, and one day, inevitably, our foe will lie bleeding to death at our feet, as the world is now."

In recent years the scale of farming and the practices employed have resulted in food production and distribution that account for a third of our greenhouse gas emissions. Agriculture through the ages can be seen as a massive, harmful geo-engineering project, albeit an unintentional one.

Should the introduction of biochar to mitigate climate change be regarded as another risky geo-engineering project?

Dark areas of soil found in the Amazon demonstrated the possibility of charcoal achieving a permanent improvement to infertile soil (see Chapter Three on the Amazon civilization); in this it is almost unique. There is every reason to think that similar soil can be created today through the incorporation of biochar, and can help to regenerate land that has been degraded by over-cultivation. At the same time, the bio-char would reduce atmospheric carbon dioxide. This is not geo-engineering. It is agricultural development that aims to reverse the harm done by a geo-engineering project.

Biochar development could, however, become risky if it were carried out for purely commercial, not agricultural, reasons, in the way that the exploitation of fossil fuels was left to the market. A specific danger is that this could lead to establishment of fast-growing, large-scale monocultures for the production of chemicals and fuel, with biochar only a byproduct. The outcome would be a continuation of humanity's destructive relationship with the environment.

The arguments for geo-engineering are based on genuine fears about global warming. If temperatures rise more than 2°C (3.6°F) above pre-industrial levels, we can expect catastrophic changes to the climate. If runaway warming kicks in, our species, and many others, could disappear. The dangers are so great that there is an argument even for risky interventions.

When considering each proposal, there is an initial question that should be asked. Does it stabilise or reduce the level of greenhouse gases in the atmosphere? Projects like mirrors in the sky or increasing the cloud cover may be effective in reducing temperatures for a time, but knowing human nature, the result would be to allow the concentration of greenhouse gases to increase further and cause non-climate problems such as further acidification of the oceans.

Unintended geo-engineering of this nature is already taking place. Particles from wood-burning cooking stoves and other pollution block incoming solar heat and are estimated to have reduced temperatures globally by 0.5°C (0.9°F). This is cited by scientists as a major cause of concern. It just means that we tend to ignore the real effects of greenhouse gases—for a time.

Many proposals are made to reduce the emission of gases, but fewer to reduce the amount already in the atmosphere. Both are necessary.

Here are some of the proposals made recently for the former: to reduce emissions. A direct-current electric grid would lose little energy in transmission and could replace coal, gas, and nuclear generation by connecting solar-steam generators in the desert to tidal, wind, and wave power in the North. Feed-in-tariffs are enabling people to generate their own electricity either with photovoltaic panels or by joining community-organised wind farms (an option mostly unavailable in North America at this time). Domestic fuel cells would convert energy at the location where it is to be used. The little "rocket" cooking stove in the hands of millions of poor families could reduce deforestation and also reduce the soot that falls on glaciers, where it increases the rate of melting. Fuel could be produced from farming algae. Management of grasslands could retain more carbon in the soil. A global money system that is not based on debt and interest would dispense with the financial necessity for irrational economic growth.

The second part—extracting carbon dioxide from the air—is more difficult, and biochar may provide the primary technology. Spreading lime on the oceans is a proposal that could possibly be used. The lime would draw carbonic acid from the water and enable it to suck extra carbon dioxide from the air. But it has not been tested, and may not be affordable or even adequate to counter the carbon emissions resulting from refining the lime.

Better—particularly organic—farming practice would result in more carbon being extracted from the air and kept in the soil for as long as the good practice continues.

The use of biochar, when linked to good farming practice, has the additional advantage of locking carbon in the soil permanently. The greatest potential is for doing this while restoring degraded land. It may be the only technology capable of extracting sufficient carbon dioxide from the atmosphere to save us from the worst effects of global warming. It is a natural process that restores the balance of carbon between soil and atmosphere. And it has been tested and is immediately available. It seems to be a unique technology for application throughout the world. Therefore, it should be adopted as policy by the United Nations, by international meetings of leaders such as the G8 and G24, and by national governments.

Downsides

All suggestions for modifying the climate need to be investigated for possible harm. It is worth reflecting that nearly all the major interventions that have contributed to our modern standard of living have had a downside.

The climate crisis is a direct result of the Industrial Revolution and subsequent exploitation of fossil fuels that brought so many benefits. Nuclear science, that energy that promised to be "too cheap to meter," continues to trick governments into writing blank checks, and will almost certainly leave future generations with radioactive waste and power stations they may not have the expertise to manage. It has also provided weapons of mass destruction that, as yet, have only been used against civilian populations by a "responsible" nation. Water-based sewage revolutionized urban health, but is a one-way process for transporting essential nutrients out to sea, and is also a major contributor to the global shortage of clean water. Agrichemicals, which were an essential part of the Green Revolution, leave biologically dead soil. If the present rate of fishing using advanced technology continues, the oceans will be effectively dead by 2050.

There are other threats for the future due to the technologies on which we rely; for example, our dependence on interconnected electricity grids. Solar activity has been calm for decades, but a violent solar storm is predicted for 2012. If this matched the Carrington event of 1859, it could destroy worldwide electronic communications. And what about the computers and embedded chips on which almost all aspects of modern life are dependent? A virus attack could cause chaos. The utopia of nuclear fusion would result in dangerously extreme centralization of energy production. We don't yet know whether genetic engineering will or will not cause disruption to plant and animal life on a global scale. In May 2009 the American Academy of Environmental Medicine (AAEM) called for an immediate moratorium on GM foods, saying, "GM foods pose a serious health risk in the areas of toxicology, allergy and immune function, reproductive health, and metabolic, physiologic and genetic health."

If biochar is to be used on such a scale as to influence climatic conditions, it must be classed as a major intervention alongside, and on a similar scale to, the interventions mentioned above. Most of these have started by providing benefit but have in the end resulted in serious hazard. The lesson is that biochar application should be regulated in order to be restorative—reversing the harm we have done to the atmosphere and the soil—and carefully monitored to identify potential hazards. As I have said above, it should not be determined by market forces.

Predictions

Since I first started writing about these issues in the first edition of *The Little Earth Book* in 2000, I have learned to ignore predictions. Predictions produce targets that are mostly inadequate and rarely met. They allow politicians, public authorities, and industrialists to ignore principle. And climate events have consistently turned out to be worse than predicted.

Predictions are constantly being made about the effect of greenhouse gas concentrations on temperature and the climate. It is generally accepted that global temperatures must not exceed 2°C (3.6°F) above pre-industrial levels if we are to avoid catastrophic consequences. It is becoming questionable whether we will achieve this, so there is talk among politicians of limiting the figure to 3°C (5.4°F), for no scientific reason.

Most pernicious is the theory that emissions caused by industry or individuals (usually in the West) don't matter if they are "offset" by paying money for reductions elsewhere (usually in the global South). This was adopted as part of the Kyoto Protocol, and is the basis of flourishing profit-making trade. Biochar could greatly increase the credits available for further emissions of greenhouse gas by industry.

This Briefing will not side with any of the various predictions about how much we should reduce emissions, or by when. I will only mention my personal inclination. I listen to Rajendra Pachauri, head of the Intergovernmental Panel on Climate Change (IPCC), who is advising that the 2°C (3.6°F) target should be reduced to 1.5°C (2.7°F), not increased. And to James Hansen, heading the NASA Goddard Institute for Space Studies, who advises that concentrations of carbon dioxide in the atmosphere need to be brought down rapidly from the present 387ppm to 350ppm, and subsequently right down to pre-industrial levels of 280ppm. The IPCC says that this can only be achieved if by 2050 rich countries reduce emissions by 95 percent and all others by 85 percent; if there is a complete ending of deforestation; and if ways are found to sequester carbon dioxide. This means that there is no capacity for soil modified with biochar to serve as an offset for emissions. Consequently we cannot have a system of carbon credits to service such offsets. There should be separate and independent control systems: one to reduce emissions from burning fossil fuels and the other to maintain carbon in the land, and add to it.

Scientists have now moved on from presenting the science in a dispassionate way to expressing urgent personal concern—"shouting from the rooftops," as Al Gore puts it. A typical example was an editorial in *New Scientist* in 2005: "With green groups playing politics, scientists seem to stand alone. In recent months they have reported compelling evidence that climate change is a real and present danger, and that the global climate system may be on the brink of dangerous positive feedbacks. At this magazine we regularly meet climate and earth-system scientists who harbour real fears for themselves and their families about what the 21st century will bring. Jim Hansen, George Bush's top climate modeller, said last week that we have 'at most 10 years to make drastic cuts in emissions that might head off climate convulsions.'" That was five years ago, and during that time emissions have been rising. For ten years then, read five years now.

Their predictions have since become even more extreme, and events like the release of methane from permafrost—that could easily induce a feedback where more warming causes more releases which cause more warming, more releases, more warming—have developed much faster than previously predicted. Even if we were to stop emitting carbon dioxide today, the world would still be getting hotter for several decades. We may be approaching a tipping point where it will be impossible to prevent the feedback effects that make further warming unstoppable. When writing about these things it is not easy to appear "balanced" and dispassionate, nor to avoid repetition. They need to be said again and again.

And charcoal? Again there are various predictions. Some say that it is the only proven way to extract carbon dioxide from the atmosphere. For Chris Goodall it is one out of "ten technologies to save the planet," eight of which would reduce emissions. Only two of his ten technologies would extract existing greenhouse gas: biochar is one, and the proper management of soils and forests is the other.

The technology is too young to be dogmatic. It has immense possibilities and it has dangers if driven by unregulated commerce. Its possibilities, however, indicate that it should be taken very seriously indeed. Governments should do all they can to finance research, subsidize equipment, and encourage farmers to experiment.

Chapter 3

The Amazon Civilization

This is what we are taught: the Amazon region has been a virgin rainforest since the beginning of time, inhabited by hunter-gatherer tribes isolated from civilization. Perhaps history needs to be rewritten.

Try this. The Amazon region was the site of one of the great agrarian civilizations of the world—with charcoal fundamental to its sustainable agriculture—that lasted from the time of Plato until a pandemic destroyed it 500 years ago. The forest took over with greater diversity than previously. The isolated tribes we meet today are the progeny of traumatised survivors.

Archaeologists have found it easier to study dry savannas than the dark depths of rainforests. Only in the last 30 years has evidence gradually come to light: first in the Bolivian headwaters, then on the large Marajo island where the Amazon flows into the Atlantic, then in much of the deepest interior.

The Spaniards, in the 15th century, had heard stories of the Chibcha people of Colombia in the headwaters of the great river system, who rolled their chieftain in gold and washed him ceremonially in a sacred

lake. This gave rise to the legend of El Dorado and the hunt for fabulous wealth. Added to this were stories of spices, such as had sparked Dutch and British rivalry in the East Indies.

An account of Francisco de Orellana's epic journey still exists. He was probably a relative of Pizarro and participated as a child in the conquest of Peru, where he picked up a number of

Indian languages, so Pizarro chose him at the age of 30 to lead an expedition looking for gold and spices downstream. It was one of the most surprisingly successful expeditions in history and he was the first European to travel down the full length of the river from Ecuador to the Atlantic. He encountered a tribe of women warriors that reminded him of the Amazons of Greek mythology, so he adopted the name for them.

Here is a diary entry from the time when he arrived at the junction with the Rio Negro, halfway down: "There was one town that stretched for fifteen miles without any space from house to house, which was a marvellous thing to behold. There were many roads here that entered into the interior of the land, very fine highways. Inland from the river to a distance of six miles more or less, there could be seen some very large cities that glistened in white and, besides this, the land is as fertile and as normal in appearance as our Spain." The river and its tributaries were the long-distance highways.

Fred Pearce quotes archaeologists finding tens of thousands of miles of raised banks across the Bolivian Amazon. By corrugating the flooded fields, farmers created ridges on which they could plant their crops clear of the floodwaters, with the furrows collecting water for the dry season. The digging and earth-moving involved in creating these structures, says Clark Erickson, the archaeologist who found them, is "comparable to building the pyramids. They completely altered the landscape." Charcoal buried in the mounds has been analysed and suggests that they were created more than 2,000 years ago. Erickson also found possibly 193 square miles (500 km²) of fishponds and weirs. He says that grassland would be flooded and stocked in the rainy season, after which people would transfer the fish to ponds for extracting in the dry season. This comment fascinated me, because Richard St George had previously told me about monastic fish farming in Europe that followed exactly the same pattern. In France I found a large level field beside a monastery with controllable channels for filling and draining it from and to the adjacent stream, together with a formal tank for keeping the fish through the winter.

Another archaeologist, Heckenberger, focused on nineteen settlements about two and a half miles (4 km) apart in central Brazil, each on a raised area about one mile (2 km) long, linked by a series of boulevards up to fifty-four yards (50 m) across. In the surrounding swampy land there were bridges, dams, dykes, causeways, canals, ponds, gardens, orchards and places for medicinal plants. At the heart of each town was a big circular plaza from which roads radiated. He describes it as a garden city. He dates the area he studied to between 1200 and 1500AD.

What interested me are the things not found. In this and other descriptions there is no mention of pyramids as in the Maya civilization, no ramparts, no hierarchy of grand buildings surrounded by hovels. Is it too much to speculate that the abundant fertility did away with the need for a highly centralized authoritarian society?

It all disappeared. The population was decimated by European diseases to which the Indians had no resistance. One can only imagine the effect of this pandemic. The farmers, shopkeepers, metalworkers, priests, scholars, and their families, who must have made up these societies, fled into the jungle to become hunters and gatherers. Nine out of ten died. The forest took over. Later travellers could find little trace of the civilization, and assumed that Francisco de Orellana had deliberately misled them with his stories of El Dorado.

Dark soil (terra preta)

In 1871 pockets of deep rich dark soil were discovered along the waterways, though its origin was not known. One can find various references to it since then. In 1928 the soil was described as naturally fertile. In 1941 it was thought that these soils might be the sediment of old lakes or that they might consist of volcanic ash. The locals call it *terra preta*. Only in the 1950s, 60s and 70s was the idea of a human origin gradually accepted.

As I mentioned in the introduction, Amazon soil generally is thin, red, acidic, and infertile. The *terra preta* is alkaline, and is characterised by the presence of low-temperature charcoal in high concentrations. It includes plant residues, animal feces, fish and animal bones, and other organic material, along with a large number of pottery shards. It is rich in nutrients such as nitrogen (N), phosphorus (P), calcium (Ca), zinc

(Zn) and manganese (Mn). It also shows high levels of microorganism activity. It is not prone to nutrient leaching.

Soil, presumably, was mixed with a mulch of organic waste and charcoal. The mulch permeated the black soils with microorganisms—bacteria and fungi—that still ensure it regenerates itself even when used for cultivation. Bruno Glaser of the University of Bayreuth says, "The really important point here is that the soils contain charred residues. That is different from the residues of [natural] burning. Both of them improve soil fertility, but [burnt] residues don't last for long, while [charred] residues have a long-term effect on soil fertility, acting over centuries. It's at least as good as manure. In some places we know that Indians successfully farmed land containing black soil for 2,500 years or more."

Photographs of the excavations I have seen show no sign of interruption or layering of the kind you would expect in other societies due to wars or economic changes. So we could be talking about a continuous civilization that lasted from the great days of ancient Greece up to the time when the Conquistadores brought a pandemic from Europe.

The important point for this Briefing is the contrast between the Amazon and other civilizations whose agriculture was their downfall: Mesopotamia and North Africa (the breadbasket of Rome) come to mind as spectacular cases. Egypt was one of the few that bucked the trend, but this was due to the annual inundation of silt from the Nile—until the Aswan Dam prevented it, making its fertility dependent on hydrocarbon-derived chemicals. It was charcoal incorporated into the soil that enabled the Amazon civilization's agriculture to remain fertile and productive through the centuries.

The discovery of this civilization sparked the recent interest in biochar as an ingredient for sustainable farming. Not only did it provide fertile ground among infertile surroundings, but the soil it left unused for centuries remains fertile today. Local farmers have long used *terra preta*, dug out and sold in bags, for its amazing ability to revive exhausted soils. When spread on fields it retains its fertile qualities for long periods. It even seems to reproduce itself: Amazonian people say that *terra preta* can lose fertility if it is over-cultivated, just like other soils, but if it is left undisturbed for a few years its original qualities will return.

Lehmann gives various references to charcoal being used as a soil-enhancer from that period up to the present. A 17th-century Japanese text on agriculture refers to "fire manure"—almost certainly biochar. W. H. Trimble in the US in 1851 noticed "evidence upon almost every farm in the country in which I live, of the effect of charcoal dust in increasing and quickening vegetation." In 1878 there were reports from China of black earth that contained charcoal improving plant vigor. In the 1920s, J. Morley wrote an article on compost and charcoal, and observed that "charcoal acts as a sponge in the soil, absorbing and retaining water, gases and solutions"; and "as a purifier of the soil and an absorber of moisture, charcoal has no equal." In the 1930s charcoal was being marketed for turf applications. E. H. Tryon in 1948 recorded detailed scientific information on "the effect of charcoal on certain physical, chemical and biological properties of forest soils." In Japan, biochar research intensified in the 1980s.

History suggests that charcoal could be fundamental to the future of agriculture.

Chapter 4

Biochar and Agriculture

*"The main danger to the soil, and therewith not only to agriculture
but to civilisation as a whole, stems from the townsman's
determination to apply to agriculture the principles of industry."*

−*Fritz Schumacher*, Small is Beautiful, 1973

One of my favorite statements is by Herb Stein, an adviser to President
Nixon who either thought there is no point in taking action to avert the
inevitable or, alternatively, had a rather low opinion of politicians'
intelligence. "If something can't go on forever," he said, "it will probably stop." Industrial farming falls into this category, yet the government
and its advisers seem desperate to extend present policies into the
future as if nothing will change.

We are entering a period of immense changes. Some are obvious,
some are possible but uncertain, and some, particularly those relating
to climate, may fall into Donald Rumsfeld's famous category of
"unknown unknowns." When thinking about the introduction of biochar into farming we should concentrate on farming practices that will
last rather than on those that will need to be abandoned.

Global warming, in addition to its other dangers, is now acknowledged as a serious threat to the production of food. And the production
of food is one of its greatest causes. "Agriculture, forestry and other
changes in land use are responsible for 30 percent of human-caused
greenhouse gas emissions," says a *Worldwatch Report* in 2009. "Changing
how we grow crops, raise livestock and use land can reduce greenhouse
gas emissions and increase carbon sequestration and storage." In order to

achieve a stable climate, these emissions don't just need to be reduced to zero; the land needs to become a net sink for carbon dioxide. The IPCC says that better agricultural practice could extract at least 4.4 billion tons of carbon dioxide equivalent from the atmosphere annually by 2030.

Much can be done with an understanding of how carbon is retained in the soil through good land management. Biochar would be a further bonus: it extracts yet more carbon from the atmosphere; it results in carbon being retained more permanently; it increases fertility; it is effective in reviving degraded soil; and is of particular value in the tropics, where temperatures above 25°C (77°F) increase the rate at which soil organic matter is oxidized.

Much of the literature on biochar applies the new technology to existing farming practice. The first half of this chapter, therefore, is devoted to the reasons why many aspects of industrial farming cannot be regarded as permanent. Organic farming, at the other extreme, is often considered a matter of opinion, prejudice, or lifestyle choice. But we are facing both a climate crisis and a food crisis, and the discussion should not be polarized. We need analysis of how all agriculture can be made sustainable and how biochar can contribute to this.

The transition to sustainable cultivation will make profound changes to our lives. It will affect how and by whom food is grown. It will affect distribution: where and from whom we buy our food. And it will open up new possibilities for where we live and what occupations we pursue. On the psychological side, the transition will lead to new ideas about how we relate to each other, what we mean by security, and where we look for satisfactions. Integrated into these changes will be the use of biochar to reduce greenhouse gases and increase yield. Some communities are already looking for ways to adjust to the changing situation through the Transition movement, as described in Rob Hopkins' book, *The Transition Handbook*.

Biochar should be regarded as a natural component of sustainable farming. The book *Biochar for Environmental Management* gives evidence that most soils contain some char that was left by forest fires during the last few thousand years. In some places there are significant quantities. Biochar, therefore, is not an alien introduction like synthetic chemicals, but can be thought of as "natural," to run alongside compost, green manure, and crop rotation. Soil that has been degraded by synthetic chemicals, however, is "unnatural." The addition of biochar may benefit almost any soil if used intelligently, but it is particularly useful where the land has been compacted or lost its fertility. The book also gives evidence that biochar worked into the soil can increase soil carbon in a way that is more stable than can be achieved with ordinary compost and manures, and to support this it mentions specific studies carried out in Australia, the US, Germany, Russia, and Kenya.

The development of agriculture is partly influenced by government and partly by economics. Government and modern economics take food for granted, since, financially speaking the agricultural sector in wealthy countries is such a small part of the overall economy. This is short-sighted. Previous generations have managed without cars, central heating, television, mobile phones, and most of those things by which we measure the wealth of a nation. But they never managed without food. Now, globally, there are more people hungry than ever before. Between 2005 and 2008 the global price of wheat and corn tripled, and the carry-over stocks fell to just 61 days of global consumption, a record low. Ocean fish are being pursued to extinction. Global warming is beginning to cause droughts and floods that affect food production. Aid to poor countries is being withheld due to the recession. On top of all this comes the most immediate threat to food supplies—peak oil— and the even more serious longer-term threat of peak phosphorus. Politicians and economists will be forced to recognize the centrality of sustainable farming in the economy of the real world.

For example, the attempt to keep Britain self-sufficient in food was abandoned by the free-market policies of Margaret Thatcher and New Labour. Apples from New Zealand are cheaper than home-grown apples, so orchards were pulled up. Britain now imports 90 percent of its fruit and 47 percent of its vegetables. The uncertainty of fuel for transport and the vagaries of international finance put our imports at risk.

Threats from the agricultural base of society are not new phenomena. Classic examples, among many, are the Indus civilization and those of the Fertile Crescent, which rose and collapsed due to dependence on irrigation that brought salts to the surface. Their once-rich land has remained infertile through the centuries. Modern farming has found ways to damage the land more thoroughly and more extensively than ever before, through replacing natural nutrition with synthetic chemicals. The fate of those civilisations could be ours, not due to conflict or economic collapse, but due to failure of the agricultural base.

Soil

"Among material resources, the greatest, unquestionably, is the land." This is how Fritz Schumacher opened a chapter entitled "The Proper Use of the Land" in *Small is Beautiful*. He continued: "Study how a society uses the land and you can come to pretty reliable conclusions as to what its future will be. The land carries the topsoil and the topsoil carries an immense variety of living beings including man."

Under most conditions it takes between 3,000 and 12,000 years to build enough soil to form productive land. Soil that is richest in minerals originated from glaciers scraping the surface off rocks, and from ash and lava that have come out of volcanoes. Like most natural things, soil can recover from damage, over-grazing, or too widespread use. But it can also be lost. Degradation, compaction, erosion and salination are widespread due to bad farming practice. Soil is the largest carbon sink over which we have control.

Modern agriculture uses land as if it were an inert material on which we provide all the fertility the plants need. Artificial fertilizers were introduced in 1909 when the Haber-Bosch process produced ammonia (NH_3). This is regarded as one of the great achievements of modern science. To start with, the process just required masses of energy to extract nitrogen from air, then ways were found to get it from natural gas, and China is now using coal to produce it. But the manufacture of these fertilizers alone still uses more than one percent of global energy, so the process itself has significance for climate change.

But the use of nitrogen fertilizer produced by this process is of greater concern because it results in emissions of nitrous oxide (N_2O) that have a greenhouse effect 300 times as strong as carbon dioxide. Worldwide, the emissions of nitrous oxide caused by nitrogen fertilizers is equivalent to seven percent of all emissions from burning fossil fuels. Since industrial farming "can't go on forever" without nitrogen fertilizer, "it will probably stop."

But emissions are not the only problem. Plants don't take up all the nitrogen. Some of it gets into the groundwater or runs off the surface to pollute rivers. "Diffuse nitrate pollution puts a question mark over the future compatibility of UK food production and public water supplies," said Professor Bradley of the environmental agency ADAS in 2006. "What's more, this problem isn't limited to the UK but applies across much of Europe. The only way to safeguard the future of our water resource is to convert much of our arable land into unfertilised, restorative grassland or forest." In other words, we can either have industrially farmed food or drinking water, but not both.

Fungicides, insecticides, and herbicides now maintain the productivity of crops. They are made from fossil fuels. These -cides (dictionary definition: a thing that kills) have a disturbing origin. Chemical companies that developed poison gases for the two world wars and for Vietnam needed to have something to sell in peacetime, so they modified their chemicals to kill insects and microbes rather than people and forests. Rachel Carson wrote *Silent Spring* in 1962 to warn of the consequences, which (we now see) are contributing to the fifth mass extinction: a "microbial" extinction with similarities to four of the other ones. But the attack on the land has only intensified. The companies are now developing herbicide- and insecticide-tolerant genetically modified (GM) plants so that more poisons can be sprayed onto food crops right up to the time of harvest, which is hardly reassuring for the consumer. The companies are, in effect, carrying out a massive experiment on us and on the land with virtually no research into the long-term consequences. In Britain the Food Standards Agency seems to be running an aggressive campaign to denigrate organic farming and support the use of synthetic chemicals and GM crops.

Livestock corporations find it more efficient—in terms of profit—to separate animals from crops, depriving animals of a natural environment. This makes two problems out of one solution. The solution was to mix straw and dung to fertilize the fields. The two problems are how to dispose of the excrement and what to do with the straw. One fears that biochar may be introduced into the process to justify the status quo.

The Green Revolution, which resulted from the development of a few thirsty hybrid plants dependent on artificial chemicals, was immensely successful. These crops had minimum contact with the soil through diminished roots, and put all their energy into the grain. As a result modern agriculture enabled the population to increase exponentially. The recommendation for birth control that Rev. Thomas Malthus made in 1798, and for which he was stigmatized, was ignored until recently.

But this approach to farming destroyed the natural factory of the soil: without biological structure to hold it together, it drifts into the oceans or blows away in the wind. It has been estimated that the US has lost half its topsoil during the last 100 years. The arid cornfields of northern France and East Anglia have probably suffered in the same way. This is an aspect of cultivation where biochar could be useful due to its moisture-retention properties and its affinity with the roots of plants.

Industrial farming in many locations relies on irrigation fed by deep-bore wells that drain ancient "fossil" aquifers. Some, like the Ogallala under America's wheat belt or the one under Arizona, have only a decade or two left.

There is widespread concern over the future of bees for pollinating crops. A third of our food is dependent on their services. Colony Collapse Disorder (CCD) is now widely reported. In the UK a third of beehives have been lost over the last two years. No single reason has been identified, but it must be at least partly related to monoculture crops and our use of synthetic chemicals throughout the countryside.

It has been known for the last ten years that a particularly nasty group of insecticides called neonicotinoids has been directly responsible for the loss of hives. They work by blocking specific neural pathways in the insects' central nervous system, and this prevents forager bees from imparting precise directions to the others. Neonicotinoids were banned in France, Germany and Italy, though the US, Canada, and UK still allow them. The UK has carried out no research, and uses the lack of proof as a reason not to resist pressure from corporate lobbies. We will, of course, be saved the cost of research when there are no bees left on which to do research. But we will be hungry.

The plow has been with us since the dawn of agriculture, so one can't blame industrial farming for all its faults. However, it turns soil upside down, suffocates aerobic microbes and exposes anaerobic microbes to oxygen. Soil carbon, also exposed, oxidizes into carbon dioxide. The soil becomes less fertile and greenhouse gases are released. It is more damaging in temperatures over 25°C (77°F). Loss of carbon—often referred to as loss of soil organic matter (SOM)—is the reason why it has been essential to regenerate the land with manure, compost, green manure, and crop rotation. The deterioration could be contained when plows were pulled by horses, but 100 horse-power tractors have massively accelerated the damage, and are one of the reasons why chemical fertilizers have become essential. Think of the postcards of idyllic rural scenes with birds flocking behind the horse-drawn plow, then watch a modern plow in action: not a bird in sight. The birds know that the soil is dead.

Nearly 250 million acres, seven percent of the world's arable land, is under no-till management. The amount is growing rapidly as rising fossil fuel prices increase the cost of tillage and also to reduce topsoil losses. The practice sometimes uses nitrogen fertilizer drilled in with genetically modified seed that allow herbicides to control the weeds but this approach retains many of the problems of industrial farming. In Parana, Brazil, however, farmers have developed no-till organic systems. The farmers have found that the yield of wheat and soybeans is a third more than conventionally plowed plots. It has reduced labour and fossil fuel costs, and has improved soil biodiversity. The practice is spreading, and researchers around the US, including at the Rodale Institute in Pennsylvania, are also developing organic no-till tools and methods.

Sustainable farming and gardening methods are being tried, but as yet they are regarded as marginal, incidental to the serious business of industrial farming. However, when industrial farming collapses we will be left with chemical-free methods such as organic cultivation, permaculture, forest gardens, allotments, and backyard gardens to provide the bulk of our food. Academic and practical research into biochar needs to focus primarily on these sustainable farming methods.

Productivity

In the absence of synthetic enhancements, we will need to make maximum use of productive land.

Contrary to common perception, industrial farming is remarkably inefficient in terms of the amount of land used. It is only relatively productive as regards the small number of people employed. Mixed (animal and crop) small-scale farming is much more productive in terms of output per acre.

The most productive use of land is a carefully managed vegetable garden or allotment feeding a family and neighbours. As the size goes up, to market gardens and small mixed farms that supply shops, the amount of waste increases and the productivity per acre goes down. Large-scale industrial farms, feeding supermarkets that require standardized products, are responsible for a huge amount of waste. What's more, this kind of farming causes a significant proportion of human-induced greenhouse gas emissions through its mechanization, transport, fertilizers, and nitrous oxide emissions.

A report by the University of Reading in June 2009 entitled *England and Wales under organic agriculture* compared the output from the two main branches of agriculture: conventional and organic farming. It found that organic fruit and vegetable yields compare favorably with conventional agriculture. Chicken, egg, and pig meat would fall to roughly a quarter of current levels due to the abandonment of animal factories. This would mean less grain consumed by animals so the amount of cereals consumed by humans would be maintained. Beef and lamb production would increase to double its present level. The report draws attention to additional benefits: there would be less water pollution, synthetic fertilizer inputs would cease, wildlife would benefit, jobs in the countryside would increase, greenhouse gas emissions would reduce and, above all, the soil would have an increased carbon content.

Biochar for Environmental Management has many examples and scientific studies that show biochar providing an increase in the yield of crops in both tropical and temperate areas.

During World War II, when labor and other resources for agriculture were allocated to the military, the Victory Garden program produced a remarkable increase in productivity. It is the informal sector where an increase in food production could occur.

If the government were convinced that sustainable farming could actually increase the amount of food produced, the argument would then turn towards the number of workers available for work on the land. Economists say that, in industrialized countries, our prosperity depends on the majority of the population designing things, working in manufacturing or service industries, or in the financial sector. They sometimes admit that farming in poor countries needs less mechanization. But a surprising statistic shows that the number of people in both rich and poor countries getting food from field to mouth is not very different: between 40 and 60 percent of the population. In industrialized countries these people are working as grocery clerks, in distribution centers, drawing up contracts, driving to supermarkets and a host of other activities. Count the tractor trailers you see on the highway and you will find that two out of five are transporting food from one place to another. Food handling—except for the people who actually cultivate the produce—is one of the most profitable businesses in the country.

To get the full benefit of high productivity per acre, more people would need to work on the land and more produce would need to be sold in local markets. Large farms will need to be divided into smaller hands-on farms where the farmer is more aware of what is happening in the soil than on the balance sheet. For many of us this will provide a new focus for our lives.

Schumacher argued that reconciliation of people with the natural world is no longer merely desirable—it has become a necessity. "Instead of searching for means to accelerate the drift out of agriculture," he said, "we should be searching for policies to reconstruct rural culture, to open the land for gainful occupation to larger numbers of people, whether it be on a full-time or a part-time basis, and to orientate all our actions on the land towards the threefold ideal of health, beauty and permanence."

Peak oil

"We are close to this turning point, a sort of turning point for mankind, when this critical energy, for agriculture in particular which means food and people, is heading down."
–Colin Campbell, co-founder of the Association for the Study of Peak Oil and Gas, speaking in "A Farm for the Future"

The availability of oil and all products that are dependent on it will start a steep decline fairly soon. The actual date is not important. The decline is inevitable whether it starts in two years or ten years.

The 100-year climb in the production of oil levelled off in 2005 at about 74 million barrels a day. It peaked at 74.8mb/d in July 2008, then rapidly fell to 71mb/d in May 2009 (keeping in mind that the declining economy played a role in the severity of the change). The independent website The Oil Drum says the output is likely to slowly decline until 2012 and then plunge 3.4 percent a year. But there are other reasons for concern besides the exhaustion of oil fields. Conflict could result in a sudden, if temporary, drop; or global agreement on the need to control emissions and the amount of oil extracted could force a reduction in the supply.

At present virtually all crop fertilization, food production, and food distribution is dependent on oil or natural gas. A farmer relies on a 100hp tractor, and a distributor relies on 500bhp trucks. It would be facetious to suggest that you will see 100 horses pulling a plow across a field or 500 horses dragging a large cart down the motorway, but the image helps us appreciate our dependence on cheap energy.

The amount of cereal produced globally will decline, whether we like it or not, due to peak oil. But, since about half the world's cereal harvest goes to feed animals, there is an obvious need for the consumption of meat to reduce. To synthesise 1 lb of meat, for example, a cow consumes 20 lb of plant protein, so eating grain is much more efficient than eating animals fed on grain. Another calculation adds to the argument due to water shortage: it takes eight times as much water to produce a pound of beef as a pound of vegetables. Regional differences come in: grass-fed cattle in temperate climates are sustainable, but this is very different from cattle ranches that require rainforests to be converted to soybean plantations. The future may be in having fewer animals and allowing them to eat the plant material they evolved to eat, fertilizing the ground with their droppings at the same time. Yet another cause of concern is that the meat and dairy industry produces a significant proportion of our greenhouse gas emissions.

Oil has enabled us to develop systems that work well provided there is sufficient fuel and fertilizer, provided the wizardry of electronic accounting gets food from farm (anywhere in the world) to shelf when needed, provided no one in the amazingly interdependent distribution process goes on strike, provided the global economy and trade practices are sufficiently stable to allow us to import much of our food, provided etc., etc. But a break in this complex chain could happen at any time. Most food in rich countries is sold through supermarkets that have no storage space of their own, so they are dependent on just-in-time deliveries from huge distribution centers. A strike by haulers could result in empty shelves within days. The system in Britain nearly collapsed in 1979 with the truckers' strike, and again with the fuel crisis of 2000. In a sustainable system, the cultivation and consumption of food would need to be brought close to each other.

Cuba was forced to face the problems that we will all face when its supply of oil was suddenly curtailed in 1991, following the collapse of its "sponsor," the Soviet Union. In just a few years it changed from the intensive use of oil in a centralized system to dispersed, smallholder, largely organic, production. Most of Havana's food is now grown in private plots of less than one-third acre and sold from stalls outside the growers' homes, at street corners and under the covered walkways of its elegant, crumbling, colonial buildings.

There is not much sign that other fuels will ever adequately replace oil to enable us to continue using tractors and trucks in the way we do at present. Farming and the transfer of food from field to plate will have to undergo a revolution. But agriculture, as defined in neoliberal analysis, represents only two percent of national wealth so the necessary changes do not show on the radar of governments—yet.

Peak phosphorus

"Quite simply, without phosphorus we cannot produce food. Phosphorus is as critical for all modern economies as water."

–Dana Cordell, Institute for Sustainable Futures, Sydney, Australia

Phosphorus is essential to all living things and has no synthetic alternative. In our bodies, for example, adenosine triphosphate is a nucleotide phosphorus molecule found in every single cell, and it drives the thousands of biological processes needed to sustain growth, muscle contraction, movement, and reproduction.

Peak oil could be disastrous for modern agriculture. Peak phosphorus could be worse. In just over a year, the price of phosphate rock has surged more than 700 percent. China has reserves but is discouraging export in order to protect its farmers. This concerns both India and Europe because they are totally dependent on imports. Production in the USA has dropped 20 percent in the last three years due to lack of sources, and it has started to import phosphorus from Morocco.

In medieval times Europe got its phosphorus from pigeons, and later from the 23-foot (7m) deep guano on Pacific islands. Nauru became the richest country in the world per capita, until it was left looking like a spoil heap. Guano wars broke out, and the US gave its citizens the right to claim ownership of any islands that could supply the stuff. Then rock phosphate took over. Morocco holds a third of the world's proven reserves. We can expect the emergence of a small number of "phosphate superpowers" to rival OPEC's control over crude oil.

The demand for phosphorus is growing. Crop-based biofuels are cornering it in the agricultural system in unprecedented volumes. And there is a new threat from the nuclear industry: it says that, though there are only 4.5 million tons of conventional high-grade uranium to be mined, there are 20 million tons in rock phosphate deposits. Now the US military has started to use phosphorus to dissolve its enemies. These facts blow holes in the concept of both biofuels and uranium being regarded as sustainable sources of energy.

Research at Newcastle University shows that without phosphates the yield of wheat could plummet from the present 4t/a (tons per acre) to as low as 1.8t/a by 2040, and thereafter drop to 1.1t/a. It says, however, that organic wheat, grown without chemical fertilizers, yields 2.7t/a. Richard Young of the Soil Association, who is also an organic farmer specializing in cattle, told me: "We urgently need to find a safe way to recycle phosphorus. One wicked thing in this respect is that bone meal is not allowed on an agricultural scale due to BSE ["mad cow disease"], and we [organic farmers]—who have never had, nor ever will have a case of BSE—are forced by the government to pay £25 ($40) a week to have the bones from our butchers shop taken away and incinerated, instead of being allowed to return them to our own land as we did before 1996." He adds, "Once we run out of phosphorus, yields from intensive agriculture will fall by at least 50 percent, offering the possibility of mass starvation, even if the population doesn't increase."

Most of the phosphorus consumed by humans is secreted in urine, and subsequently flushed out to sea. So sewerage systems should be changed. A project in Ouagadougou, the capital of Burkina Faso, found that urine placed in opaque containers becomes free of bacteria in a month, and dry faecal matter in six months. The project is spreading because it is cheaper than traditional sewage collection and treatment, and it provides valuable fertilizer. The construction of toilets that separate urine from feces has become a new local industry.

Among rich countries only Sweden is pioneering ecological sanitation with its "Closing the Loop" program. This involves Ecosan toilets that separate urine at source. In Britain, we should avoid the lazy but expensive option of replacing crumbling Victorian sewers in which feces, urine, and rainwater are mixed—thus generating huge treatment costs, and a huge waste of resource—and make the thorough change to an ecological system.

My friend Dr. Ravikumar, who developed the Anila biochar stove in south India, is also developing toilets that separate urine at source. In *The Big Earth Book* I wrote about similar developments by Dr. Samer Kurvey at Wardha, Madhya Pradesh. He said that if the use of urine as fertilizer became standard practice throughout India, the country would have no need for synthetic fertilizers.

There is a natural resistance to putting urine directly on food crops, and this is where biochar has an important role. Biochar needs to be activated with nutrients before it is used in the soil, otherwise it will spend the first year absorbing microbes and nutrients from the surrounding soil, and crop yields will fall not rise. Urine is one of the best mediums to "charge" biochar. One solution is to have pits where biochar is mixed with compost, manure and urine. The mixture will then be incorporated in soil and the biochar will add permanence to the retention of nutrients in addition to the sequestration of carbon.

Pandemics

"There has been a transition from old-fashioned pig pens to vast excremental hells, containing tens of thousands of animals with weakened immune systems suffocating in heat and manure while exchanging pathogens at blinding velocity with their fellow inmates."

—Mike Davis, professor of history UCLA, 2009

Factory farms reduce animals' lives to misery, corrupt their genetic inheritance and provide an ideal breeding-ground for pathogens—all in the pursuit of profit. The World Organisation for Animal Health says that three-quarters of recent emerging diseases originate in animals. "Big pharma" won the battle to commercialize antibiotics, with the result that they have been excessively used as prophylactics. They are now losing their effectiveness as resistant bacteria develop. This will lead to more intensive use and more resistant bacteria. Already, just in the UK, more than 200,000 people are affected annually by serious antibiotic-resistant infection, and there are approximately 1,500 deaths from MRSA, 9,000 deaths from *Clostridium difficile* and 4,000 deaths from ESBL *E. coli*.

Bird flu caused a panic in 2007 because it killed half of those infected, but luckily it did not become a pandemic as feared. However, a mutation could develop. Genetic Resources Action International said in 2006, a year before it hit Britain: "The deadly H5N1 strain of bird flu is essentially a problem of industrial poultry practices. Its epicentre is the factory farms of China and Southeast Asia and . . . its main vector is the transnational poultry industry, which sends the products and waste of its farms around the world." In spite of this warning no action was taken. The first outbreak in Britain was in a well-managed factory farm with the highest bio-safety standards in the country: it could not have been caught from a passing swan, as the industry likes to suggest. This demonstrates how viruses can spread around the world in a matter of weeks due to current animal management.

Genetically selected fast-growing pigs with weakened immune systems, kept in filthy and crowded conditions, provide hundreds of opportunities for a virus to mutate as it jumps between them. The US Center for Disease Control says that the current swine flu outbreak is similar to the viruses that have been circulating in American pig farms since the 1990s. Pigs are genetically closer to humans than chickens and are therefore more likely to produce the variants to which we are most susceptible. Swine flu is aptly named: it has been called the pigs' revenge on the swine that keep them in such appalling conditions.

A new strain of MRSA (Methicillin-resistant,Staphylococcus aureus), the superbug resistant to most antibiotics, is spreading in farm animals on the continent. It is responsible for a quarter of all MRSA cases in Dutch hospitals and is starting to appear in the UK. The government and its advisers seem only concerned to reassure people that things are "under control." They have not been prepared to challenge the hugely powerful food industry or to face up to the implications of adopting sustainable animal husbandry, let alone the possibility of taking morality into account with the treatment of animals. Their only consideration seems to be the fear that the cost of meat would go up and the supply would go down.

Many biochar enthusiasts identify the waste from these animal factories, particularly those managing pigs and chickens, as a major source of biochar feedstock. The waste is rich in nutrients; it is immediately available in large quantities; and its use as feedstock for making biochar would replace the cost of disposal with an opportunity for profit. There is little doubt that this will be one of the most immediate areas for the development of biochar technology. But it would encourage the perpetuation of an obscene and dangerous practice. The backlash could bring the use of biochar in other situations into disrepute.

The demise of the otherwise sustainable Amazon civilization may be an extreme example of the potential effect of a pandemic, but it provides a compelling reason to question some modern agricultural procedures.

Introduction of biochar

This may have seemed a long diversion from the discussion of biochar. The purpose is to indicate which aspects of farming have a long-term future and which do not. Food scarcity is likely to increase, so in order to produce the maximum amount of food from the available land, research and development will have to be directed towards farming and horticulture that has maximum yield. The success of the Amazonian civilization suggests that biochar should have a major role in sustainable organic farming. Its chief claims are the ability to stabilize nutrients in the soil, loosen compacted and heavy soils, give surfaces for microorganisms that transform nutrients, and retain moisture, though this is not an exhaustive list.

If you look at any piece of charcoal, even biochar dust, it is obvious that it has the potential to store moisture. Biochar retains much of the structure of the plants from which it is made. This means that it is riddled with microscopic holes that initially repel water—the reason for its being light in weight—but in due course in the soil it becomes saturated and stabilized by mycorrhizae on plant roots. This ability to retain moisture is one of its most important attributes for agriculture. It is certainly the one that first appealed to the farmers we talked to in south India.

In poor countries the rising price of chemical fertilizers is already causing severe hardship. Debt is forcing many off the land, and there is little chance of them finding employment elsewhere. In India, over half the population is dependent on the land and is able to feed itself largely

Scanning Electron Microscope (SEM)
photographs of biochar

Agapanthus flower head before pyrolysis

Agapanthus flower head
after pyrolysis
in an Anila stove.

The plant structure
remains intact when it is
turned into charcoal.

3mm   x15

Enlargement taken from
the photo above.

The biochar appears white
in these SEM photographs
because of the imaging
process—it is of course
carbon and therefore black.

500um   x 80

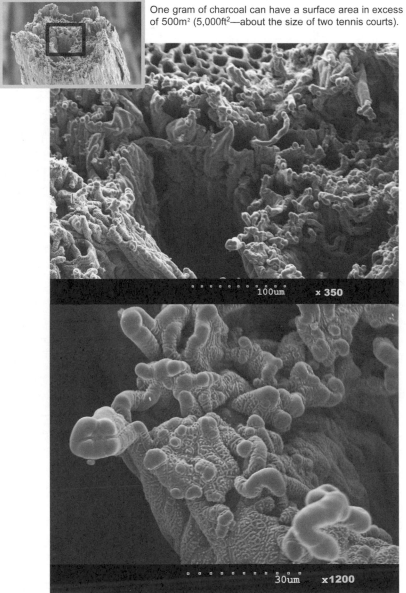

One gram of charcoal can have a surface area in excess of 500m² (5,000ft²—about the size of two tennis courts).

with its own produce, so it is vitally important that smallholder farms are viable and part of a marketing network. My experience in south India, as well as from reading about many pilot schemes, convinces me that research and development should primarily concentrate on helping small-scale mixed—cattle and crop—farms to benefit from biochar. It is these farmers that control much of the world's productive land. Large industrial schemes grab the headlines, but a billion smallholders would have a greater effect in reducing greenhouse gases if the use of biochar were to become a standard part of farming practice throughout the world.

A joint report by the UN Food and Agriculture Organization (FAO) and the World Bank in June 2009 says that an additional 988 million acres, straddling 25 African countries, are suitable for farming. It refers to models in Thailand where degraded land had been transformed by smallholder farmers. As in Thailand, it says, future success will come by using agriculture to lift Africa's smallholder farmers out of poverty, aided by strong government measures to guarantee their rights to land.

According to another report issued in 2009 by the FAO and the Organisation for Economic Co-operation Development (OECD), an extra 4 billion acres can be added to the existing 3.5 billion acres of cropland in the world. Like the above OECD report, this also says that the way forward depends on better land rights for smallholder farmers.

If this is correct, there is ample opportunity to grow grasses and fast-growing trees specifically to produce biochar, provided this is integrated into farms or for the supply of neighboring farms. The FAO emphasis on smallholders reinforces the argument against monoculture plantations for agribusiness as a separate operation from farming.

Biochar can be made from any organic material, whether it is wood, leaves, straw, food scraps, or even from sewage. Some of these materials may simply be wasted if not converted into biochar. In Western society there is a huge amount of wasted organic material that is expensive to deal with and that causes damage or pollution. On the other hand there may be competition from other needs—compost, for example—so biochar would need to be integrated into existing practices. Crops may be specially grown for making biochar, though if this is done to excess monocultures could result and the production of food could suffer.

Stylized arrangement of a typical smallholder village in India.

Commentators have tended to simplify the potential feedstock down to just a few sources and talked about the associated benefits or dangers. Anyone who has travelled widely will appreciate that the variety of source can be endless, and can vary with different cultures and in different ecosystems.

The Worldwatch Institute says that simply using waste materials for biochar—such as forest thinnings, rice husks, groundnut shells, and urban waste—could sequester 661 million tons of carbon dioxide equivalent per year, and "far more could be generated by planting and converting trees." I referred on page 39 to Professor Bradley's comment about nitrate pollution: "The only way to safeguard the future of our water resource is to convert much of our arable land into unfertilised, restorative grassland or forest." This is just one example of the great variety of opportunities to achieve feedstock for biochar.

With well-managed mixed and organic farming there is virtually no waste. All biomass is recycled to retain as many of the nutrients as possible, and rotation allows carbon and nitrogen to be captured from the air. If biochar is to be introduced into the organic process it will only be done if the additional, particularly the long-term, benefits are understood. For example, biochar mixed with compost or manure can make the microbes and nutrients more stable than they would be in compost or manure alone. The Amazon *terra preta* demonstrates the long-term benefits of biochar for soil enhancement that are not found elsewhere.

As I mentioned in the Introduction, there is a need for an organization to advise and certify good practice, in the way that standards for organic farming are certified.

I have mentioned some problems of feedstock from industrial farming. Municipal green urban waste is a more positive source for turning into biochar, and it can be distributed locally for gardens and allotments. Then there are food and crop processing facilities, wood and paper processing industries, construction and demolition materials. It helps when these provide a source on a single location because there is less bulk to be transported in the biochar than in the feedstock. However, many sources from urban and industrial areas will need to be critically examined as they may contain unacceptable amounts of heavy metals.

Trees retain the carbon in their trunks, branches and roots for decades, so nothing is gained by carbonizing established trees until the end of their lives. The exception, of course, is with fast-growing trees and coppices, but these can be regarded as an aspect of farming. Annual plants and leaves or cuttings from trees only retain the carbon for a matter of months, so by carbonizing them, carbon is captured.

Where, and how much?

In July 2008, the FAO issued a database on the world's soils. It includes the "global Carbon Gap Map that allows the identification of areas where soil carbon storage is greatest and the physical potential for billions of tons of additional carbon to be sequestered in degraded soils." This is of critical importance for establishing policies for the application of biochar on a large scale, keeping in mind the FAO's recommendation that this should be done largely by smallholder farmers.

The FAO says that until now most efforts have involved above ground sequestration, primarily through planting trees, but soils are the largest reservoir of the terrestrial carbon cycle. Depending on how it is used, soil can be a sink or a source of greenhouse gases. "For long-term sequestration," it says, "organic carbon must be stored in forms and in locations in the soil profile with slow turnover." The atmosphere contains less than half the carbon that is stored in the soil, so "relatively small changes in the flow of carbon into or out of soils have a significant effect on a global scale."

In addition, NASA Goddard Space Flight Center and others (SEDAC) have produced a map of net primary productivity (NPP), which it measures in units of elemental carbon, representing the primary food energy source for the world's ecosystems. SEDAC also issued a map of human appropriation of this resource—HANPP—which it defined as "the amount of carbon required to derive food and fibre products consumed by humans."

Craig Sams, founder of Green and Black's organic chocolate and a former chairman of the Soil Association, has given a vivid statistic: if all productive land were devoted to producing biochar for just one year, then enough carbon could be sequestered to take atmospheric concentrations back to pre-industrial levels. Admittedly, we'd all starve. On a more serious note he says that we would need just 2.5 percent of the world's productive land devoted to producing biochar to bring concentrations down to pre-industrial levels by 2050. He is now putting his recommendation into practice (with Dan Morrell of Future Forests) in their company Carbon Gold Ltd, setting up a series of community-scale biochar projects around the world including Belize, Mozambique, Brazil, the Maldives, the US, and Europe. If the FAO is right about the potential increase in the amount of productive land that could be achieved, and if Craig Sams is right in his calculations, then the use of just 2.5 percent devoted to biochar is entirely feasible and could bring concentrations of carbon dioxide in the atmosphere down to safe levels—provided, of course, that emissions are reduced as well. This could form the basis of global policy.

And how much biochar should be added to land in order to affect fertility? Tests, where significant increases in crop yield have been recorded, have used between one-half and seven tons per acre, a wide range. Many used 2t/a, though some as little as 0.1t/a. The amount needed should obviously be determined by analysis of the local conditions or by trial and error. There are some cases, however, where yields have decreased due either to biochar being applied to excessively alkaline soils or due to the biochar being fresh and "uncharged."

Permanence in soil

Biochar should not be considered an inert material that remains unaltered in the soil. To start with, its main contribution is to increase the retention of moisture. It attracts nutrients and provides surfaces on which microbes attach themselves. Mycorrhizal hairs linked to roots explore the cavities, and cation exchange allows minerals to be taken up. In time the biochar is aggregated and becomes fully a part of the soil. The biochar one puts in now will be enhancing the soil for the next 50 to 100 years, possibly longer.

Initially it is insects and small mammals that fragment it. Later, as described above, the changes are largely chemical and microbial. One study found particles up to 20mm across were covered and penetrated with fine plant roots after only a month. In another the root hyphae had broken down the lumps.

Carbon will usually oxidize in the soil over time, but how long? The carbon in *terra preta* seems effectively permanent, and renews itself. Oxidation can often take thousands of years. Radiocarbon dating has identified carbon residues from forest fires that took place more than 10,000 years ago. Other studies from coastal temperate rainforest of western Vancouver suggest an average half-life of 7,000 years. In northern Australian woodlands there is carbon that is 700 to 9,000 years old. In the dry conditions of northern Australia an age of 1,300 to 2,600 years is normal. A ten-year study of biochar in mesh bags buried in a temperate forest found no measurable decay at all. But occasionally shorter residence periods are reported. The turnover time for charcoal from fires in the Russian steppes, for example, is possibly 300 years.

In contrast to the long residence periods frequently found, other studies found relatively quick loss often for reasons other than oxidation. Fresh charcoal can simply float away in a flood due to its low density before being saturated or being colonized by mycorrhizae.

Fresh biochar is light and, if spread on the surface, could be blown away to join other aerosols like fine soot from wood-burning cooking stoves that shield the planet from solar energy and give the false impression that emissions are not so serious. They also land on snow, darken the surface, and increase the rate that glaciers melt. So surface dressing is only appropriate if the biochar can be protected from the wind and covered to prevent oxidation.

But surface dressing may be necessary, for example with no-till cultivation. In these cases it can be mixed with compost, distributed with liquid manure, covered with plant material or formed into pellets. In woods it can be distributed where leaf cover will soon protect it. Earthworms and burrowing rodents move biochar downward to the level containing plant roots. Biochar plowed into soil, for example, has been found at levels lower than the plow depth. Research at the sites of forest fires has also found charcoal taken deep into the subsoil. Fungi play a key role in changing biochar properties, and particularly in distributing its components laterally.

Wood-burning cooking stove giving off soot.

Care has to be taken over storage and distribution in areas vulnerable to fire as it could obviously ignite. Charcoal dust is a hazard to health and is, no doubt, the reason why traditional charcoal burners have a shorter life than their peers. The danger varies with the type of feedstock used. This is of particular importance for home-based village schemes in poor areas that cannot be subject to controls.

Many people in Britain would like to have a small pyrolysis unit, like the Anila stove, for their garden or allotment. I fear that this might be an enthusiasm that would wane as the tedium of managing it kicks in. It might be more appropriate for garden centers to sell bags of biochar mixed with compost that are ready for use to enhance soil, or for local authorities to distribute it having prepared the activated biochar from urban waste. Group allotments and areas of woodland could have permanent brick-built Adam retorts that are easy to fill with waste organic material and left to smolder for a day.

Chapter 5

Pilot Schemes

"Evaluation does not rely on a fundamental advance in science, but on the application and adaptation of existing science."

—Johannes Lehmann, Testimony to US House
Select Committee, June 2009

Many pilot schemes are in progress. I will start by describing some with which David Friese-Greene and I have contact, and then move on to others that I feel have particular significance.

The banana grower

Pattu Murugeshan has been using biochar for four years.

Pattu Murugeshan has been farming at Kootambuli in Tuticorin District, Tamil Nadu, following in the footsteps of his father, grandfather, and great-grandfather. They have all alternately grown one year of rice paddy, from which they can get two or three crops, followed by two years of bananas. David Friese-Greene was making inquiries about charcoal in the area and came to hear about Murugeshan, so he arranged to visit the farm, with Rob Bryant of Swansea University, to get the details. Friese-Greene and Bryant carried out this research but they emphasize that figures and details are based on hearsay and Murugeshan's recollection, not on verifiable measurement. What is certain is that Murugeshan is convinced of the benefit, and the practice is spreading.

The bananas go into freshly dug holes on a six-and-a-half-foot (2m) square grid, giving 1,000 plants to the acre. After three months men dig pits and arrange drip irrigation while women apply urea, phosphate, and potash fertilizers carefully to each plant, working five hours a day from 8am to 1pm.

Four years ago, when taking his rice to the mill, Murugeshan found 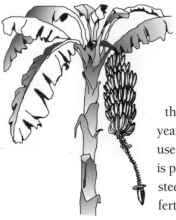 that the mill had piles of rice husk charcoal as a by-product. He decided to add it to his plants. He tried an even spread but came to the conclusion it was better to work 2.2 lb or 4.4 lb into each pit, thus using one to two tons per acre each banana season. He says that the char can be identified the following year so the effect is cumulative. As a result his use of fertilizer has dropped by a quarter, which is particularly important because its cost is rising steeply. Only two, not three, applications of the fertilizers are now necessary, so labor costs have gone down as well. Water used to be provided for two hours a day; this has now been reduced to one hour a day. Each plant produces 29 lb of bananas whereas previously it only produced 20 lb: a 44 percent increase. The bananas ripen more slowly after picking, so he can send them to Kerala or Chennai and get a better price. And, he says, the bananas taste better.

Ten others in their Farmers Association have already adopted his practice. The other twelve initially resisted the idea because it seemed to introduce an additional process—more work!—but they are likely to follow suit.

However this means that there will not be adequate charcoal at the rice mill to meet their needs and, due to added demand, the price will go up. At present he buys the rice husk charcoal at 2 cents/lb. Commercial charcoal made from wood costs 8 cents/lb but it would have to be ground to powder before use, which is a costly and hazardous process. The farmers now have a strong incentive to produce fine-grained charcoal—biochar—themselves.

Banana leaves have many uses. Some are left to cover the ground, keep it moist, and rot in due course. They are used as plates in restaurants and subsequently eaten by pigs. The trunks, up to eight feet (2.5m) tall and rich in phosphate, also have various uses but are frequently burned. If, say, half of them—estimated as 13t/a—were used for pyrolysis they would provide 4.5t/a of biochar. After a few years, when the soil is sufficiently rich in carbon, this might give them spare biochar to sell. Alternatively the Farmers' Association could set aside a part of their land for fast-growing elephant grass to add to the banana waste in order to provide the feedstock. These figures are highly conjectural estimates from Rob and David, and would need more rigorous testing to have serious validity.

But whatever the exact figures, the farmers themselves are convinced of the benefit of incorporating biochar into the cultivation of bananas. The route taken was trial and error, followed by observed increase in yield, followed by adoption by neighboring farmers. Only then did scientists try to catch up and analyze the reasons and figures.

Murugeshan's project demonstrates the benefit that biochar can bring. It suggests that there is a need for appropriate pyrolysis units that can be integrated into the farming process. It also suggests that appropriate biochar crops can be grown as part of an integrated process without the need for buying charcoal made by hazardous traditional methods, or for the production of biochar to be carried out by enterprises that use monoculture plantations.

SCAD

"I have tried to bring social change for 25 years
and every step forward has been pushed back
by global forces beyond my control."

–Cletus Babu, in conversation, 2003

Dr. Cletus Babu and his wife Amali started SCAD (Social Change and Development) in 1986, and I have known them since 2001. Their disappointment with global institutions, free trade, and corporations that try to make farmers dependent on alien products has not dampened their drive to improve conditions in the villages of southern India.

They have introduced many environmental schemes. At an early stage students were organized to collect and scatter tree-seeds over the surrounding hillside: in time the hills turned green and changed the microclimate. Rainwater harvesting in several forms is being promoted. SCAD has a scheme for people to plant and maintain appropriate and useful trees. It encourages families to keep rotational *"char-bagh"* plots for nutritional and medicine plants next to their homes. It has programs for vermiculture and composting. It runs an organic demonstration farm to give out plants and advice. It also has many social schemes covering women's groups, salt-pan workers, leprosy, and children with mental problems. Their distribution of organic cowpea biscuits has rid the area of vitamin A deficiency in a region where night-blindness used to be endemic. The result of all this practical work is that the farmers trust SCAD and its animators.

SCAD offers a good opportunity to test the use of biochar because of its relationship with 450 villages. A village might consist of 1,500 people in 300 family groups, half of which own some land. Each village has some women's groups, with twenty women in each group. There are also animators who keep them in touch with the organization.

When I talked to Cletus and Amali in January 2008 about the growing interest in biochar, they immediately understood the concept and its global significance: if millions of farming families were involved, it might be possible to reduce the amount of wood needed for cooking, soil could be improved, and burial of the biochar would help to counter global warming. It could also provide a source of income in villages to help stem migration to cities.

David Friese-Greene has visited SCAD several times to work with it on the project and coordinate with developments taking place within the International Biochar Initiative. In due course SCAD invited Dr. Ravikumar to test a pyrolysis cooking stove he had been developing during the previous eight years for use in villages.

SCAD's enthusiasm for the project also gave the opportunity for Matt Dunwell and Reinhart von Zschock, who have experience with similar issues in the UK, to get involved with the needs of communities in a poor country, and for scientists in the universities of Swansea and Bristol to carry out analysis.

Utra Mankasingh of Bristol University said that soil in the area is predominantly red and laterite with low moisture content and very low soil organic matter (less than 1.2 percent SOM). Agronomists consider soils with less than 1.7 percent SOM to be in the pre-desertification stage, and tropical soils lose SOM quicker than temperate soils. Chemicals have degraded it, particularly nitrogenous fertilizer that is subsidized by the government. The monsoon is no longer reliable. Rains leach nutrients from the soil. Some farmers speak of there being no birds any more. The government is reducing its subsidy for fertilizers, and the aim is for charged biochar together with nitrogen-fixing plants to increase water retention, SOM levels, the fertility and, of course, to sequester carbon dioxide.

The ability of biochar to retain moisture appealed to the farmers immediately. Other aspects, like improving the yield of crops were of interest but would have to be demonstrated. So trial plots were initiated.

Some of the test beds showed a definite increase in yield, though a few showed a slight reduction. For good results it was found that the amount applied should be above 3t/a (tons per acre), so the trials are ongoing. As I have noted elsewhere, when biochar is fresh it can attract nutrients and microbes from the surrounding soil, and the plants suffer for the first year or two. For this reason the biochar is being charged with manure and compost in specially prepared pits. Urine is perhaps the best nutrient. One of Dr. Ravikumar's many interests is the introduction of toilets that separate urine from feces and, if adopted as a sewerage scheme, this should enable it to be added to the biochar pits. The mixture could then be worked into fields.

In another test, biochar is being immersed in "oorani" catchment depressions together with compost material beside wells, to be charged over a period of about six months while monsoon rainwater filters through subsoil into the wells.

Ravikumar's cooking stove, called the Anila stove, can be fed with any plant material and produces good quality biochar. It is a standalone unit and prone to be knocked over by children, so Ravikumar and the team started work on modifying the design to be more akin to traditional mud-formed stoves. The amount of biochar that a family could produce in this way would be adequate for improving the soil of a family's char-bagh, with surplus being sold to a pool.

The greatest benefit of the project has been in demonstrating problems and opportunities. Various women were encouraged to use the Anila, but it was soon apparent that the inflexibility of intense heat for about an hour did not suit their cooking patterns. This seems to indicate that pyrolysis stoves are unlikely to have general application for cooking. SCAD is buying charcoal from the families with Anila stoves for preparing it with compost and selling as fertilizer. This brings a source of income into the villages.

A larger unit in each village, managed by women's groups, would be more effective: such as the ICPS (Improved Charcoal Production System, or "Adam retort") developed by Chris Adam in Germany. The feedstock would not have to be so carefully prepared, so prosopis, a pervasive thorny weed, could possibly be used if ways could be found to avoid its health hazards. The biochar would be charged with nutrients (compost, manure, and urine) before being used for soil enhancement, or it could be sold as fertilizer. The pyrolysis process produces heat, and ways are being studied to utilize this, for example to dry the next batch of feedstock or to run a Stirling motor for charging batteries.

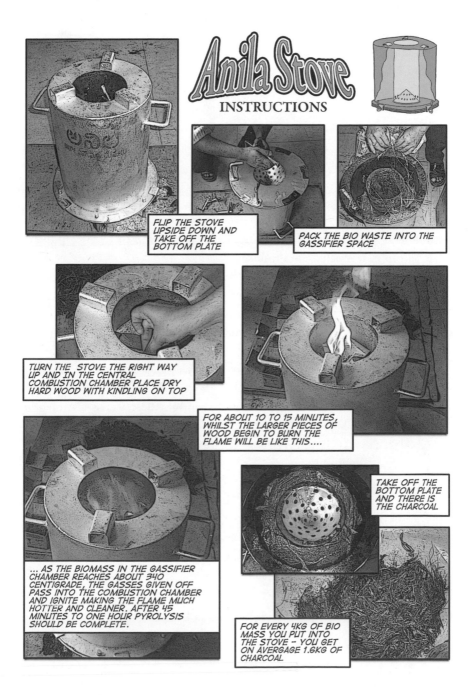

Anila Stove
INSTRUCTIONS

FLIP THE STOVE UPSIDE DOWN AND TAKE OFF THE BOTTOM PLATE

PACK THE BIO WASTE INTO THE GASSIFIER SPACE

TURN THE STOVE THE RIGHT WAY UP AND IN THE CENTRAL COMBUSTION CHAMBER PLACE DRY HARD WOOD WITH KINDLING ON TOP

FOR ABOUT 10 TO 15 MINUTES, WHILST THE LARGER PIECES OF WOOD BEGIN TO BURN THE FLAME WILL BE LIKE THIS....

... AS THE BIOMASS IN THE GASSIFIER CHAMBER REACHES ABOUT 340 CENTIGRADE, THE GASSES GIVEN OFF PASS INTO THE COMBUSTION CHAMBER AND IGNITE MAKING THE FLAME MUCH HOTTER AND CLEANER. AFTER 45 MINUTES TO ONE HOUR PYROLYSIS SHOULD BE COMPLETE.

TAKE OFF THE BOTTOM PLATE AND THERE IS THE CHARCOAL

FOR EVERY 4KG OF BIO MASS YOU PUT INTO THE STOVE – YOU GET ON AVERAGE 1.6KG OF CHARCOAL

It was recognized that handling charcoal should be kept to a minimum and carefully controlled. Charcoal dust is a major health hazard with traditional production and the less need for handling and collection the better. This would also be a problem if wood were used in larger units: it would have to be pulverized under controlled conditions.

The Adam retort.

The SCAD College says that the farmers typically produce 3.1t/a (tons per acre) of biomass. 0.4t/a of a cereal crop is grain, and some of the rest is used for cattle feed and compost. Therefore about 1.3t/a would be available for pyrolysis, giving o.4 to 0.7t/a of biochar. This is thought to be adequate for an annual top-up in order to gradually improve soil quality but more, perhaps 2.2t/a, would be required for an initial injection.

Special grasses can produce 11t/a, so might be used as an integral part of the allocation of land for cultivation; they thrive on poor soil.

Alternatively, coppice or fast-growing trees could be grown, again as an integral part of the farming operation. Fast-growing plants or trees are usually thirsty, and this could be a serious drawback in these arid conditions. Another source for biochar might be from the neighboring state of Kerala, which has ample water; perhaps it could be made from water hyacinth that contaminates all its canals.

Oil drum pyrolysis unit.

The UK team designed a low-cost horizontal pyrolysis unit that could be suitable for a women's group to manage in a village. It was subsequently made in SCAD's technical college. It didn't work too well because the metal of the inner drum was thick and reduced the transfer of heat, whereupon SCAD's engineers enthusiastically suggested and made alterations. This could be the best outcome since it meant that they "took ownership" of the design and will continue to refine it themselves.

While medium-sized village units are desirable, there appears to be a need for a larger one. David Friese-Greene is now discussing a more sophisticated pyrolysis unit designed by Gabriel Gallagher and available for construction under licence by SCAD's Technical College. This runs at 320°C (608°F) and would produce 440 tons of biochar a year on a continuous rotating basis. The heat generated by the unit would be useful for another project being discussed: a biodigester dealing with urban waste that needs to be dried and subsequently kept warm.

A local-exchange currency could be linked to the distribution of appropriate equipment for pyrolysis by women's groups in the villages. This suggestion from Richard Douthwaite, who visited SCAD recently, is based on the Liquidity Network being tested in Ireland. It would make them less dependent on earning rupees and, in addition, encourage local trade. This local exchange scheme would provide a useful fall-back should the national currency experience problems similar to those in the West.

As can be seen from the above, the production and use of biochar at SCAD is a process that has only recently started. There is great enthusiasm from the organization, but its ongoing development will depend entirely on being able to demonstrate to farmers that it will increase the quality of their thin soil and the yield of crops. I find it of particular interest because of the holistic approach to the needs of a large community. Cletus Babu has good connections in India—the President visited him in 2007—so the organization's experience could be widely disseminated.

Southern France

This is not a pilot scheme but a suggestion for the introduction of pyrolysis stoves in rural Europe. It is based on wooded areas of southern France that I know well.

The Languedoc region has extensive oak and chestnut forest, much of which was previously used for grazing sheep, as is evident from dividing walls and domed shelters. Farmers now cut down patches for firewood and leave them to regrow for 20 or 30 years; in spite of this, the amount of woodland in France is increasing. One can see long stacks of logs outside most farms and hamlets. The government encourages the use of wood for heating because it is "carbon-neutral"—carbon dioxide emitted when the wood burns had been captured from the air as the original tree grew and is recaptured by new tree growth. The government pays you half the cost for installing an efficient modern wood-burning stove.

It is the gases from wood that provide heat. When making biochar, pyrolysis—heating the wood in the absence of oxygen—provides heat as the gases are expelled and burn, while leaving the carbon structure of the wood intact. If the charcoal is buried, the process becomes

"carbon-negative" because the carbon dioxide is not returned to the air. This would provide an incentive for the government to subsidize pyrolysis stoves instead of wood-burning stoves.

The heat needs of these villages and farmers could be researched, and appropriate stoves developed. Wood would be heated in the absence of oxygen. Some of the heat would be used to start the next batch. The charcoal residue from the stove would be crushed for use as biochar on the land.

I have not heard of this being done yet, but most adaptations of the pyrolysis process are in very early stages. This would be just one opportunity out of many for technical development and start-up businesses. Farmers would benefit if it can be demonstrated that biochar added to the soil can increase fertility. This area of France is subject to periodic droughts, as with the heat wave of 2003, so biochar's ability to increase the retention of moisture would be appreciated. Alternatively, the farmer could mix the biochar with compost and manure and sell it.

Carbon Gold in Belize

I have referred to this project in the chapter on agriculture (page 57).

Green and Black's chocolates have been an ethical choice for people with a sweet tooth. Cultivation of the cacao trees in the Toledo district of Belize, from which the chocolate was made, resulted in a lot of prunings that Craig Sams recognized could be used as feedstock for the production of biochar. Additional material could be received from neighbouring farmers who practiced shifting cultivation that resulted in logs and sticks being burned. Other sources were also identified, such as waste wood from sawmills, sawdust, crowns, and branches. This provided a basis for the project.

Three different types of pyrolysis units are being used. There are simple drum kilns that can be relocated to places where crops have been harvested. There is the Adam retort, which I describe below. Then there are Adam retorts working in series, where higher volumes of biomass are available. In addition, continuous cycle retorts are in the process of development.

The Adam retort was chosen because it can be built in about one week at low cost—possibly the equivalent of $500—with locally available blocks or bricks. The chambers are large so it can be easily loaded with material that does not have to be cut into small pieces

Initially the firebox is lit using any combustible material for about 30 minutes. This is enough to get the biomass to smolder, giving off gases that are led into the firebox and keep it at about 450°C (842°F) for up to 30 hours. It is then unloaded and filled, allowing three burns a week. It was initially developed in Auroville, India, as an alternative to traditional mud-covered mounds for making charcoal; these gave off methane, caused health problems, and produced less than fifteen percent weight of charcoal from the weight of wood used. The Adam retort takes about one ton of feedstock per burn and produces a third of a ton of biochar. It makes sense to locate the retort close to places where the feedstock grows.

Carbon Gold is working with the Toledo Cacao Growers Association (TCGA), which involves 1,200 farmers. Under current practice, their residues are either burned or left to decay. The TCGA is subject to fair-trade rules, which allows Carbon Gold to conduct strict monitoring and verification. It is also working with two timber organizations that operate sustainable forestry activities but have a lot of waste material. All the biochar from these projects will be used to produce organic fertilizer to be sold in Belize.

Ghana

*This scheme demonstrates the need for secure land-holding.
It is reported in* Biochar for Environmental Management.

Asuano is a small village in Ghana with a population of 760, and is connected to a town 20 miles away by a gravel road. Three-quarters of the produce is used for domestic consumption, and the rest is sold. The sandy cultivated land is bordered by rainforest and savannah. Two crops are grown each year to coincide with the rainy seasons. All work is based on hand-tools. Saran Sohi and Edward Yeboah, the researchers who visited the village, said that land division, lack of secure ownership, and diminishing fertility are the main problems.

The farmers are not interested in carbon sequestration; they are only concerned with the yield of their crops. But carbon sequestration would be achieved as a by-product if charcoal were used to increase the yield from their land.

Faustina Addai is not a typical farmer in the village because she has secure ownership of twelve acres. She has been using charcoal to improve the yield of her crops for 20 years.

Since all farming is based on hand-tools, continuous cultivation of such a large area is not possible. She farms her land in two parts. Half is left to regenerate for five years. The small trees and bushes are then cut and piled up, covered in soil and partially burned: you could refer to this as "slash-and-char"—it is a traditional method for producing charcoal, which is labor-intensive and has health problems. The char is then spread over the entire land. She augments these periodic inputs of charcoal with annual charring of crop residues.

She says that her yield has gradually increased and is now double that of neighboring farmers. In her view the main reason for success is due to the increased moisture retention that the char gives to drought-susceptible sandy soils. The researchers are convinced however that her productivity would not have been maintained if there were not benefit from the nutrient-retention and regeneration qualities of biochar.

One would expect her procedures to be copied by her neighbors, but this is not the case. The researchers found that the reason was socio-economic. Most of the land in the village is rented for periods of only one or two years, providing no incentive for investment in the soil.

This is a vivid illustration of the recommendation I noted on page 54. The FAO, the World Bank, and the OECD all say that strong government measures are necessary to guarantee land rights to farmers if they are to have the chance to rise out of poverty. If the other farmers in the village had secure ownership they would follow Addai's example. The result would be a doubling of yield from the entire village without the need to expand into surrounding forested areas.

Traditional charcoal-making is a health hazard, and it is unlikely that the farmers would improve on it unaided. So there is a need for appropriate equipment to be developed and made available to them at an affordable price for producing the biochar. This could be the subject of direct government action in Ghana, or it could be funded by foreign aid.

The researchers add a comment at the end of the report, almost as an afterthought and without any analysis, that the farmers would benefit if a regime of carbon credits were introduced. To me the scheme demonstrates precisely the reverse: that carbon credits are not necessary as an incentive for these smallholder farmers to adopt the application of biochar. The need is for land rights.

Carbon credits would be an entirely new transnational mechanism that would take years to set up. It would be dependent on checking by an army of monitors. There is no possibility that this remote village would be properly served by it. And the potential for "unforeseen consequences" would be enormous. The high value of the credits would probably result in agribusiness taking over and destroying the surrounding forest, the villagers might be displaced, and the government monitors would have huge potential for corruption, resulting in social disharmony.

E. F. Schumacher explained at length that the needs of poor rural areas are seldom for financial aid but rather for knowledge and appropriate equipment. More than ever, we would be advised to keep this insight foremost in mind.

Poultry farm

This is a revealing pilot scheme that raises the dilemma posed by animal farms. It is reported in Biochar for Environmental Management.

A poultry farm in West Virginia houses 100,000 chickens with seven breeding cycles per year. The litter is fed into a pyrolysis unit (fixed-bed gasifier) operating at 500°C (932°F) to produce heat for the three poultry houses. It does away with the need for propane gas, thus saving $66,000 a year. A heat exchanger used in conjunction with the unit produces dry air; as a result, the birds increase in weight and have a higher survival rate, thus increasing their financial value. There is no longer a problem with the disposal of waste since it can be used to make high quality biochar, another financial benefit.

The biochar is a by-product, and its carbon content is dependent on the moisture content of the litter: the dryer the litter, the higher proportion of carbon. The biochar is rich in phosphorus and potassium and has an intrinsic fertilizer value in addition to its value as a soil conditioner. It is sold at $435 per ton for soybean and hay cultivation—a financial bonus. The close proximity of biochar production, energy consumption, and use on the farm keeps transport costs low, another financial benefit. The farmer says that it replaces the need for any phosphate or potassium, and there is a 20 percent reduction in the need for nitrogen fertilizer—more financial savings. Lehmann comments: "Carbon trading has not been explored in this scenario." If introduced it would add yet more financial benefit.

However, so many financial advantages are demonstrated by the project that there is no need for carbon credits to provide yet more incentive for biochar to be produced. This scheme and the experience of Faustina Addai indicate that carbon credits are not necessary either for an industrially sophisticated project in West Virginia or for farms in a remote Ghanaian village. Both projects demonstrate James Lovelock's contention, relating to the need for farmers throughout the world to bury charcoal, that ". . . this scheme would need no subsidy: the farmer would make a profit."

Factories of this kind, however, take no account of extreme cruelty to animals. They are morally indefensible. But if legislators and industrialists are impervious to ethics, then issues of health may be persuasive. The close proximity of unhealthy animals, dependent on regular doses of antibiotics and kept in tight proximity, provides an ideal breeding ground for diseases like MRSA and the 2007 avian H5N1 flu. The H1N1 swine flu, identified in April 2009, originated in similar conditions imposed on pigs.

If left to the market, the financial benefits of the biochar process linked to animal factories will simply perpetuate a cruel and dangerous aspect of industrial farming that should be phased out as quickly as possible.

Cameroon

The Biochar Fund developed a wonderfully simple cooking stove. One clay pot is filled with dry plant material and inverted into a larger pot. Some twigs are place between the two and burnt. This starts the pyrolysis process so that the escaping gas keeps the flame going. The narrow crack between the two pots allows gases to escape when the biomass is heated, but restricts access for air. A meal can be cooked and char can be emptied out when the pots have cooled. This may not be very efficient, but every village has a potter so it costs little and can be made locally. Heat from pyrolysis means that less wood is needed for cooking and the biochar can gradually enrich the family's vegetable plot.

In Cameroon, the Biochar Fund and Key Farmers Cameroon ran a pilot project to test the effectiveness of biochar. After only six weeks— and after the dedicated effort of around 1,500 small-scale farmers—the first results were available in May 2009. On some of the poorest soils, biochar shows an amazing effect: the plots with char produced four times the growth of control plots without it. In some test fields, the corn on the char plots had already begun to tassel, whereas the plants on the control were barely past the eight-leaf stage. Their biochar was produced in village-scale units from materials like palm fronds, cassava stems, weeds, and wood. This produces heat and electricity as well as the char, so they call it "combined heat and power and char." The project is moving the farmers on to slash-and-char, and away from slash-and-burn where the soil soon becomes degraded. About 400 million people in the tropics rely on slash-and-burn agriculture so this is a process that can have widely beneficial application.

Microwave

A microwave for charcoal is one idea being promoted.

When Chris Turney was a teenager he turned a potato into pure carbon after 40 minutes in the family's microwave, incidentally destroying the equipment. Years later he recalled the incident: "When we were talking about carbon sequestration I thought maybe charcoal was the way to go." He was then professor of geography at the University of Exeter. He found that using a microwave is more efficient than normal pyrolysis, turning a higher proportion of the biomass into charcoal. He thinks that the technique could take billions of tons of carbon dioxide from the atmosphere every year.

Carbonscape has developed a process for manufacturing charcoal using microwave energy. The company has begun batch-scale production but hopes to raise capital to scale up to fully integrated continuous production. The unit could be transported on a container so that biomass can be processed on site. It would also be possible to use the technology on a large scale by combining several units.

Carbon Capture and Storage (CCS)

I include carbon capture and storage because it is sometimes described as sequestering carbon dioxide. In reality, as the name implies, it is preventing some of the emissions caused by burning fossil fuels from escaping into the air. It comes into the category of reducing emissions, not reducing the concentration of greenhouse gases. It should not be confused with biochar projects that extract carbon dioxide from the atmosphere.

The UK proposes to allow new coal-fired power stations to be built if they capture just a quarter of their carbon emissions. John Shepherd, co-author of a Royal Society report on the subject, says: "Really, it needs to be 'no new coal unless you have 90 percent emissions reductions by 2020'. That is achievable."

Chapter 6

Science

"We know more about the movement of the celestial bodies than about the soil under our feet."

-Leonardo da Vinci (1452-1519)

Charcoal has three areas of benefit. First: charcoal can be used to sequester carbon dioxide from the air. Second: biochar—finely powdered charcoal—can improve the fertility of the soil. Third: pyrolysis—the process by which charcoal is made in the absence of oxygen—can provide useful by-products. Due to the present climate crisis, the most important function is the first. The second benefit could be a question of life and death to many people, since so much productive land has been degraded by overuse, desertification, and bad farming practice. Biochar can help to restore its fertility on a permanent basis. The by-products of pyrolysis are important because peak oil will result in the loss of our primary energy source as well as most plastics. Unfortunately, it is this third benefit that commands greatest monetary value in our economy. Only strong regulatory control will prevent the drive for profit sidelining the critically important first two benefits.

Photosynthesis

Leaves are usually flat, and a tree or plant has a lot of them because their principal function is to absorb solar energy. They combine carbon from carbon dioxide in the air with water from the ground to manufacture complex sugar molecules, mostly carbohydrates (CH_2O or, to be more accurate, $C_6H_{12}O_6$), the ultimate food source for almost all life on the planet. These are found in most plants and provide the plant's physical structure as well as its chemical energy. I think this is the briefest possible description of photosynthesis.

With the carbon removed from carbon dioxide, oxygen is released for animals to breathe. Falling plant material and dung from animals are carried down to form humus in the topsoil. Hairs on the plants' roots pick up nutrients that have been transformed by microbes or ionized from minerals in the soil. In due course, half the carbon is oxidized by transpiration from plants above ground and half from microorganisms on rotting matter in the soil, sending carbon dioxide back into the air and completing the carbon cycle.

Organic molecules have carbon-hydrogen bonds and are found in all living things. They are mostly carbohydrates, proteins, lipids, and nucleic acids. Inorganic molecules are substances that don't have carbon-hydrogen bonds and are not normally found in living things. These are minerals, metals, and salts. Only green plants and some bacteria can create living matter from inorganic raw material. There is much more to it, of course, but I set out below just some of the aspects relevant to the practice of incorporating biochar into soil.

On the top and underside of a leaf is a layer of thick, tough cells called the epidermis. Its primary function is to protect the interior layers of leaf tissue. The arrangement of epidermal cells determines the leaf's surface texture. A waxy layer on the epidermis, called cutin, protects the leaf from dehydration and disease. Changes in weather and light open and close epidermal "guard cells," which regulate the passage of water, oxy-

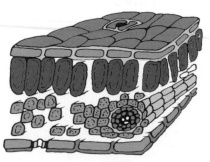

Cross-section through a leaf.

gen, and carbon dioxide into and out of the leaf through tiny openings called stomata. In most species, the majority of stomata are located on the underside of leaves. Conditions that would cause plants to lose a lot of water, such as high temperature or low humidity, stimulate these cells to close, but they remain open in mild weather. Guard cells also close in the absence of light. Look at a leaf closely and it just seems like a flat sheet of plastic. In fact, like other plant material, it is formed with a mass of microbial spaces.

Charred plant material retains the microscopic structure of plants, even leaves. This is why biochar provides cavities for plant roots to explore, and why it is so good at retaining moisture. I think of this when putting a leaf of mint in hot water to make mint tea. To start with the leaf floats. When the water starts to permeate all the invisible cavities, it sinks.

Fallen leaves and sticks cover the ground and prevent soil carbon being oxidized too quickly by the air. And the topsoil is full of life. In a handful of soil there can be a billion organisms that recycle the residues of both plants and animals. Worms and insects aerate the soil and drag rotting material down to create humus. Bacteria, fungi, protozoa, and algae release the nutrients from it in the form of minerals, proteins, carbohydrates, and sugars. Clays have an affinity to water; they shrink as they dry, allowing cavities to form and roots to explore for minerals below the topsoil. The cracks also allow biochar to descend to deeper levels where roots will be more protected from variations in temperature and humidity.

Each microscopic particle in the soil is endowed with an electrostatic charge. These particles attract positively charged ions, called cations, and repel negatively charged ones, anions. Hydrogen ions impart acidity. Countering soil acidity are the alkaline cations of sodium, potassium, calcium, and magnesium. The ion exchange process may be why soil is a cleanser: in spite of the millions of diseased bodies that have been buried, the soil does not carry their infections. In various studies the cation and anion exchange capacity of soils has increased where biochar has been added.

Soil is a storehouse from which plants extract resources and lay down reserves for the future. Mycorrhizae, the symbiotic white fungal fuzz around the roots of plants, create a living bridge between the plants, inorganic minerals, and the microbial community. It is difficult to imagine, but the surface area of these mycorrhizae can be greater than the surface area of the plants' leaves. Nutrients are taken up to nourish the plants, protecting them from pests and disease. Water infuses the soil—particularly where it is rich in humus—and is available for the plants in dry spells. As noted above, moisture retention is increased with the addition of biochar.

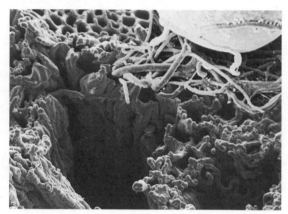

Mycorrhizae penetrating biochar cavities.

Soil is not inert mud; it has a living cyclical process of interdependent organisms and minerals. You could refer to soil as a biological factory that gets richer with time. On average the depth of soil across the globe is only about three feet (1m), incorporating topsoil of about a foot (30cm). It is this thin and vulnerable layer around the planet that makes life on Earth possible. We tend to think of conservation in terms of the physical environment, but E. O. Wilson puts it clearly in Wilson's Law: "If you save the living environment, you will automatically save the physical environment. But if you only try to save the physical environment, you will lose them both." And it is the living environment that regulates the climate.

The interdependence of all its parts also provides a warning for major interventions with biochar. Think of it like treating a sick person. Interventions may help to restore the fertility where degradation has harmed the soil. But if the interventions are too dramatic, like reducing the natural formation of humus, they could cause the patient to die. A holistic approach—integrating biochar with organic farming—is necessary, based on trials and understanding of soil structure.

The complexity of this "factory" is so immense that even today's agronomists might agree with the Leonardo da Vinci quotation at the beginning of this chapter.

Synthetic fertilizers, herbicides, pesticides and compaction by tractors of industrial farming destroy the amazing complexity of life in the soil. Without structure it cannot retain moisture and plants die with the slightest drought. Without humus, plants are deprived of nutrients other than those provided by humans. With nothing to hold it together it washes into the oceans or blows away in the wind. Industrial farming poses the sort of threat that terminated so many civilizations in the past.

Carbon dioxide

Carbon dioxide (CO_2) accounts for 78 percent of global warming.

Some figures that are frequently quoted about carbon can be confusing. The confusion is used to great effect by politicians when justifying excessively relaxed targets. The figures I give below may look daunting, but it is necessary to set them out in order to understand and check the various statements about carbon, carbon dioxide, and carbon dioxide equivalent. Also, not all scientific sources give the same figures, but this does not matter for an overview: approximations are easier to comprehend and do not seriously affect the message.

Weight

Carbon is usually measured by weight where "GtC" means a billion tons of carbon (a ton—2,000lb—is not very different from a metric ton—1,000kg). There are:

650 GtC	in plants and trees
800 GtC	in the atmosphere
3,200 GtC	in the soil (some say 1,600)
4,000 GtC	in buried fossil fuels
40,000 GtC	in the oceans

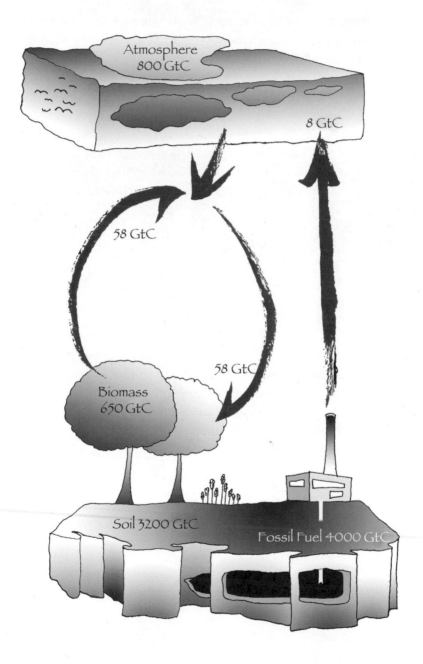

Emissions to, and concentrations in, the atmosphere are sometimes referred to as carbon (C), sometimes as carbon dioxide (CO_2) and sometimes as carbon dioxide equivalent (CO_2e); the "equivalent" includes methane and nitrous oxide. The three are not the same. One ton of carbon = 3.7 tons of carbon dioxide (the atomic weight of carbon: C~12, the atomic weight of oxygen: O~16. Add one C and two Os together: 12+32=44/12~3.7). CO_2 is usually stated as 89 percent of CO_2e calculated over a 100-year period. However methane has a much greater warming effect during its first 20 years, and a *very* much greater effect during its first five years, making it more dangerous during the coming few decades. Keeping this proviso in mind, figures for the present concentration of greenhouses gases are usually given as:

800 GtC	in the atmosphere
3,000 GtCO$_2$	in the atmosphere
3,300 GtCO$_2$e	in the atmosphere

These figures can be used to assess the amount of carbon dioxide that needs to be extracted using biochar and other techniques.

Plants capture carbon dioxide through photosynthesis, and the same amount is released from the land when microorganisms decompose matter. This is referred to as the annual carbon cycle:

58 GtC	air-earth-air carbon cycle
215 GtCO$_2$	air-earth-air carbon cycle

Each year human activity sends additional greenhouse gases into the atmosphere by burning fossil fuel and changing the use of land, where the latter provides about a third of the emissions:

8 GtC	human-induced emissions
30 GtCO$_2$	human-induced emissions
33 GtCO$_2$e	human-induced emissions

The emissions for which we are responsible (8 GtC) may seem small in relation to the amount in the atmosphere (800 GtC), but they are cumulative. Half stays in the atmosphere and some is transferred to the oceans, making them more acidic. The addition of greenhouse gases is pushing the climate system towards a tipping point where runaway global warming may take over.

Volume

However, greenhouse gases are often defined by volume: "parts per million by volume" (ppm or ppmv). So at present there are:

100ppmC	in the atmosphere
387ppmCO$_2$	in the atmosphere
434ppmCO$_2$e	in the atmosphere

The higher limit for a safe level is considered to be:

95ppmC	in the atmosphere
350ppmCO$_2$	in the atmosphere
393ppmCO$_2$e	in the atmosphere

In other words: to be reasonably safe from the worst effects of global warming it is not enough to stop adding to atmospheric carbon dioxide: we need to make a ten percent reduction from what is already there. But ultimately it should be brought down to the pre-industrial level of 280ppmCO$_2$. This shows how essential it is not only to drastically reduce emissions but also to extract carbon gases. The only proven ways to do this in the near future are through better agricultural practice, the deep burial of charcoal, and the incorporation of biochar into the soil for the benefit of plants.

Some commentators and politicians deliberately mislead by relating present levels of 387ppmCO$_2$ with a target of 450ppmCO$_2$e, without mentioning the little "e." This, they suggest, gives us time for a *gradual* reduction of emissions.

Temperature

Current global temperatures are 0.8°C (1.4°F) above pre-industrial levels. There is a time lag between greenhouse gas concentrations and their effect, so we are already committed to another 0.6°C (1.1°F). Our particle pollutants cool the planet, and when these disperse there will be a further 0.5°C (0.9°F) rise. This totals 1.9°C (3.4°F) rise in temperature even if we stopped emissions today.

It has been generally accepted that global temperatures should not be allowed to exceed 2°C (3.6°F) above pre-industrial levels. This was confirmed formally by the G8 in July 2009, though they set a totally inadequate global target of 50 percent reduction of greenhouse gas emissions by 2050. A climate science summit in March 2009 concluded: "Atmospheric CO_2 concentrations are already at levels predicted to lead to global warming of between 2°C (3.6°F) and 2.4°C (4.3°F)." These figures are a further indication that it is not enough just to reduce emissions, however drastically—we must also extract greenhouse gases from the atmosphere.

Fiona Harvey, the *Financial Times* environment correspondent, puts it like this: above the 2°C (3.6°F) level "the earth will start to experience some disastrous and irreversible damage. Agriculture would be unsustainable in many already hot regions, sea level rises would render some areas uninhabitable, and extremes of weather could have severe consequences for life and property." On top of that there is a danger that unstoppable runaway warming will kick in.

There are different views on the relationship between concentrations and the resulting temperature. The Stern Review said that above 450ppmCO_2e global temperatures have a 50 percent chance of rising to 2°C (3.6°F) above pre-industrial levels. This is only 16ppm above present levels, which are rising at 2ppm annually. So we could reach the danger zone in eight years, with a 50 percent chance of exceeding it. This shows how essential it is to start extracting carbon dioxide from the air as quickly as possible, with biochar being the only immediately available option.

Removal

The biggest stores of carbon are buried in fossil fuels and the oceans. But if we just consider the carbon cycle affecting the land, using the above figures:

> 21 percent is in plants and trees
> 27 percent is in the atmosphere
> 52 percent is in the soil

Carbon moves between the three. Each year seven percent of the carbon in the atmosphere is captured by plants and the same amount is released from the land. Therefore every fourteen years the entire volume of atmospheric carbon is recycled through plants and soil. This is referred to as "net primary productivity" (NPP).

Organic agriculture and biochar may be able to extract some of the excess carbon dioxide already in the atmosphere, but they should not be used as an excuse to relax the drive to bring emissions down to stability. Based on comments by scientists like James Hansen, the UK target of 80 percent reduction in emissions by 2050 is inadequate. We need to get to near zero carbon emissions as quickly as possible.

In addition to using biochar as a soil enhancer, charcoal can be buried deep in the ground, out of sight and out of mind. There could be scope for using some of the potential extra 3.9 billion acres of cropland identified by the FAO either for fast-growing grasses or trees for deep burial. The objection to this might be that the charcoal—or biochar—would be better used for improving the soil, assuming this extra land is infertile at present. Another objection, that it would take carbon out of the carbon cycle, can be countered with the figures given above: human emissions are small in comparison with the amount of carbon in the soil. It would be more like putting the carbon from burning coal back into the ground from whence it came.

Methane

Methane (CH_4) is a greenhouse gas with 25 times the warming effect of carbon dioxide over a 100-year period. But counting just the first 20 years it has 72 times the warming effect. And over the first—critical—ten years it is much worse. This is because methane is absorbed back to land and oceans more quickly than other greenhouse gases, so using the figure for its effect over a full century can be misleading.

Human-induced emissions of methane contribute about $1GtCO_2e$, 14 percent of anthropogenic global emissions. Globally, soils, particularly well-drained soils, are a net sink for methane. However, emissions are common from wetlands, rice paddies, and landfills.

It used to be thought that methane was more dangerous than carbon dioxide. This may still be true, because vast quantities would be released if the temperatures in northern latitudes rose above a critical point. Temperatures in the Arctic are rising considerably faster than elsewhere.

Permafrost in northern latitudes is one danger. A study by Edward Schuur of the University of Florida in 2008 doubled the previous estimates of the carbon content of permafrost to about 1,600GtC, roughly twice as much as in the atmosphere. If runaway global warming kicked in, much of this would be released as methane and there is nothing we could do to stop it. A worrying sign is that large areas of permafrost in Siberia and Alaska are already starting to melt, resulting in buckled highways and pipelines, collapsing buildings, and "drunken" forests. A researcher recently lit a candle in Siberia and jets of flame rose from the ground. This was natural gas, methane, escaping as the permafrost melted.

Methane "clathrates" are trapped in the oceans all over the world on the edge of continental shelves. These have not been included in the figures for fossil fuels. If they could be successfully exploited, they would meet all our energy needs for another century. However, the dangers are immense. Extraction of some could destabilize the rest, with huge emissions of methane and collapse of the shelves with resulting tsunamis. A decade ago, burps of clathrates on the seabed off

California raised ocean temperatures by 1°C (1.8°F). Similar burps in the Arctic could endanger the pumps that drive ocean currents (it was loss of these currents that caused the great Permian extinction, through depriving the deep oceans of oxygen). There are proposals for substituting carbon dioxide in the strata for the methane, thus sequestering the CO_2. But accidental emissions of methane would be likely. The exploitation of this immense store of carbon gas would make nonsense of attempts to work towards a zero-carbon economy, to say nothing of the hazards.

Biochar has been shown to have some effect on restraining emissions from land. One study found a complete suppression of methane emissions when biochar was applied to grassland. Another demonstrated that application of 8.9t/a on non-fertile tropical soils increased the methane sink by 0.09t/a. But not much literature on the subject exists and more research is necessary.

Nitrous oxide

Nitrous oxide (N_2O) has 300 times the warming
effect (over 100 years) of carbon dioxide,
and remains in the atmosphere almost permanently.

Human-induced emission of nitrous oxide is the equivalent of 10 percent of global emissions of carbon dioxide caused by burning fossil fuels. It is a greenhouse gas, it depletes the ozone layer in the upper atmosphere, and increases ozone at ground level. It is a very dangerous gas.

Two-thirds of all nitrous oxide emissions are caused by the use of nitrogenous fertilizer in industrial farming. Organic farming does not use this fertilizer.

Incorporating biochar into the soil has been found to reduce nitrous oxide emissions in some circumstances. In one study, where the biochar was made from municipal waste, emissions were reduced by 83 percent. In another, with chicken waste, the biochar totally prevented emissions. A three-year study with infertile tropical soils found a marked decrease in emissions of N_2O, together with increased pH and a 70 percent higher water-holding capacity. On the other hand, one project, with biochar from green waste, actually increased emissions. This shows that, while biochar can lead to significant reductions in nitrous oxide emissions, it is not necessarily true in all circumstances. Informed advice for farmers is essential.

Safe level for greenhouse gases

*"Sometimes it is not enough to do our best;
we must do what is required."*

–Sir Winston Churchill, 1874–1965

Fifty million years ago the atmosphere had a high concentration of greenhouse gases. The Earth was warm and Antarctica was free of ice. Plants and algae captured much of the carbon dioxide. The Earth was free of ice until the concentrations fell to between 550ppm and 350ppm. The concentrations then fell further so that the Earth got much colder and much of the land was covered with thick ice. During the last million years of the Ice Age carbon dioxide concentrations were around 200ppm, but there has been a series of interglacial periods during which the ice receded. In these periods the concentrations rose to around 280ppmCO_2 but never above. We have been in an interglacial period for the last 10,000 years, which allowed humanity to develop agriculture and civilizations.

Concentrations of carbon dioxide are now rising above those of the interglacial periods. They are heading back to the period when the Earth was free of ice and oceans were much higher.

Many emission reductions targets are bandied about, so who should one believe? The IPCC reports have a consensus of more than 300 climate experts, so they have great authority. But a consensus report is by definition conservative and leaves out recent research. The 2007 IPCC report, for example, when talking about sea levels, did not account for melting glaciers. The Stern review alerted politicians

because Sir Nicholas Stern is an economist and he put a *cost* against
either taking action or taking no action. He said the former would be
cheaper by far. He also included a table of probability, which is usually
ignored. It showed that we have a 50 percent chance of surviving if we
spend 1 percent of GDP on mitigation. Politicians, with remarkable
absent-mindedness, quote this as a target for expenditure, but would
you board a plane if you were told it had a 50 percent chance of crash-
ing? Anyway, he has subsequently said that his assessment of danger
was much too conservative.

I think it is more appropriate to listen to Pushkaer Kharecha and Jim
Hansen, of the NASA Godard Institute for Space Studies, since they have
access to the latest and most sophisticated information and are not con-
strained by the needs of consensus. Their 2008 report says: "Barring
prompt policy changes, [the] critical level will be passed . . . within
decades. If humanity wishes to preserve a planet similar to that on which
civilization developed and to which life on Earth is adapted, paleocli-
mate evidence and ongoing climate change suggest that CO_2 will need
to be reduced from its current 385ppm to at most 350ppm." The report
thinks that it is possible to achieve this by phasing out coal unless its
carbon dioxide emissions are *fully* (italics added) captured, and by
adopting agricultural and forestry practices that sequester carbon. It says
that the largest uncertainty in this target is the contribution of other
greenhouse gases like methane and nitrous oxide.

An increase in the concentration of greenhouse gases increases the
global temperature. As everyone knows, the weather is unpredictable
but the inertia of the huge mass of water in the oceans that expands as
its temperature rises is the best way to demonstrate the rise in the
global temperature. It is steadily rising, and 2°C (3.6°F) above pre-
industrial temperature has been widely accepted as the limit beyond
which we would face the danger of runaway global warming. At present
temperatures are 0.8°C (1.4°F) above pre-industrial levels but they are
bound to increase above 1.9°C (3.4°F) even if we immediately stopped
emissions totally.

The annual increase in emissions of greenhouse gases is usually given as a percentage above the previous year. From 1970 to 2000 global emissions increased at the rate of 1.5 percent a year. From 2001 to 2006 the annual increase was 2.1 percent. In 2007 emissions increased 2.2 percent. The failure of global agreements even to stabilize emissions, let alone reduce them, is alarming.

Since greenhouse gas concentrations are already above a figure that can be considered safe, and may have to be reduced to pre-industrial levels, we are facing a crisis. So it is not enough just to reduce emissions; the concentration of these greenhouse gases already in the atmosphere must be reduced. The potential for using charcoal to extract carbon dioxide from the air is one of the few options open to us, and the only one that is immediately available.

Biochar

Charcoal can be used for deep burial of carbon. Biochar, on the other hand, is fine-grained charcoal that is used to enhance the properties of the soil. It is made by pyrolysis: heating plant material in the absence of oxygen.

Chris Turney, a professor of geography at the University of Exeter, summarized his views like this: biochar's porous structure is ideal for trapping nutrients and beneficial microorganisms that help plants grow. It also improves drainage and can prevent up to 80 percent of greenhouse gases such as nitrous oxide and methane out of escaping from the soil.

Biochar can be made from any organic material. Its composition varies depending on the material from which it is made and the temperature of pyrolysis. It often consists of greater than 70 percent carbon but also contains small amounts of other elements (oxygen, hydrogen, nitrogen, sulphur, phosphorus, silicon, base cations, heavy metals). It gets colonised by microorganisms and fungi, and its composition changes with time in the ground.

Seen under an electron microscope, even the dust of charcoal is riddled with minute holes and passages, some smaller than others. This provides space for moisture and fungi, surfaces for microbes to cling to, and cavities for the mycorrhizae of plant roots to penetrate, along with their "helper" bacteria. Some biochar has such a fine structure that the bacteria are protected from even the smallest of soil grazers such as protozoa. Other types of biochar may just restrict larger ones like mites. The ability to exclude grazers allows soil microbes to make nutrient transformations more efficiently than when plant material that has not been charred decomposes.

Nitrogen is the single most limiting nutrient in temperate zones. In soils, the majority of nitrogen needs to be converted from locked organic nitrogen to mineral nitrogen that can be used by plants (its complex forms must be "ammonified" to NH_4+ and then "nitrified" to NO_3-). Bacteria, archaea, and fungi drive these processes. Biochar has been found to enhance these transformations in northern forests but, according to Johannes Lehmann, no evidence is yet available for agricultural soils.

Similar processes affect the mineralization of phosphorus, one of the key nutrients. But the transformation of phosphorus is also helped by the ability of biochar to increase the alkalinity of soil since this increases the amount of soluble phosphorus in the soil through cation exchanges. These allow the exchange of ions between an aqueous solution and a solid. Several studies have confirmed the enhanced uptake of phosphorus in the presence of biochar.

During pyrolysis some useful compounds can be recovered from the plant material. This can probably only be done using highly sophisticated and expensive equipment and, as I argue in this Briefing, this should not take precedence over the sequestration of carbon dioxide and the cultivation of food crops. These are some of the most common by-products.

Furfural is used either on its own or with phenol, acetone, or urea to make solid resins for use in fiberglass, some airplane components, and in vehicle brakes. It also has wide uses with solvents and processing of food.

HMF (Hydroxymethylfurfural) can be converted as a liquid biofuel that in certain ways is superior to ethanol. It can be used in the production of non-hydrocarbon plastics. And 5-HMF is being investigated as a treatment for sickle cell disease.

Levulinic acid is used in the manufacture of nylons, synthetic rubbers, plastics, and pharmaceuticals, and is a precursor in the industrial production of other chemical commodities.

Formic acid is used as a preservative and antibacterial agent in livestock feed. When sprayed on fresh hay or other silage, it arrests certain decay processes and causes the feed to retain its nutritive value longer, and so it is widely used to preserve winter feed for cattle. In the poultry industry, it is sometimes added to feed to kill salmonella bacteria. It is used to process organic latex into raw rubber. Beekeepers use formic acid as a miticide against the tracheal mite and the Varroa mite. It has many other uses: for example, fuel cells that use modified formic acid are being developed.

Nutrient qualities

While carbon's effect on global warming has been well studied, the behavior of biochar in relation to the microbial life in the soil is only recently receiving attention. Science is still catching up with observed effects.

Scientific studies have been done on wood, green waste, poultry litter, sewage sludge, straw, rice husks, coconut shells, and many other substances. The resulting biochars have a differing content of nutrients like nitrogen, phosphorus, and potassium. Poultry litter, for example, has high phosphorus content whereas some others have none. One study of crop performance attributed a seventeen percent increase in rice and a 43 percent increase in cowpea directly to the type of biochar used.

Not all the nutrients contained in the biochar are available to plants. Sewage sludge, for example, has a high content of nitrogen if pyrolyzed at a low temperature, but it is mostly organically bound. Only a small amount of mineral nitrogen is available to the plants. And it appears not to be known whether some of the less desirable chemicals from sewage are retained in the biochar and could be taken up by plants. This is of particular concern to organic farmers.

Nitrogen volatilizes at a low temperature and is lost. Phosphorus, on the other hand, does not do so until about 700°C (1,290°F). Chicken feces are particularly rich in phosphorus, but it may well be that more nutrition is retained if the feces are properly managed as manure than if pyrolyzed.

Trials have shown that nutrients contained in the biochar itself can have some benefit to plants, but usually only to a limited degree. Of greater significance is the performance of biochar once it is in the soil. Most studies attribute benefits to the indirect effect of biochar, such as the microbial life it encourages, the reduced use of fertilizers, maintaining the pH of soil, reducing the acidity of the soil, neutralizing toxins, increased moisture retention, the break-up of clods, preventing fertilizers leaching into streams, etc. There have however been cases where reduced yields resulted due to the biochar being applied to already alkaline soils, making them excessively alkaline. These studies show the need for appropriate advice to be available to farmers.

The incorporation of biochar into soil usually has a beneficial effect on the soil's fertility, and sometimes this is dramatic. But the scientific reason why this is so lags behind observation of experiments, pilot schemes, and the evidence of long-charred remains. Trial and error— the method that was obviously used to develop *terra preta* centuries ago—remains the most important need. But scientific understanding will speed the application and avoid pitfalls.

The ability of biochar to modify soil pH has an influence on the availability of various trace metals but the process is too complex to describe simply. This property is particularly useful with acidic soils. Biochar needs to be "charged" with the nutrients before incorporation into the soil, otherwise it may reduce crop yields for the first couple of years as microbes are migrating into its cavities. It is usual to mix the freshly made biochar with compost or manure before application.

Soil fertility depends on nutrients being returned to the soil.

Farmers have always appreciated this, and it is the basis of organic farming. The nutritional value of soil together with its retention of moisture can be enhanced by biochar, but it is obvious that a balance has to be achieved between the use of plant material for compost and manure and its conversion into biochar.

Chapter 7

Carbon Credits?

"The present economic crisis is partly generated by a huge overestimation of the wisdom of the market process."

–*Amartya Sen*, The New York Review of Books, March 2009

At first glance it seems reasonable that buried charcoal should earn carbon credits under the Kyoto Protocol or its successor. Charcoal is largely the carbon content of plants that have captured carbon dioxide from the air, and getting it into the ground takes carbon dioxide out of the air. Saleable credits would give you an incentive to do this.

But there are two main problems with the suggestion that biochar should be incorporated into the present cap-and-trade mechanisms of the Kyoto Protocol. First, the many biochar processes involved and assessment of their effectiveness would take years to research, define, and evaluate with sufficient precision to provide an acceptable basis for global trade. Second, the economic framework in which trading would take place has entered a period of extreme unpredictability.

An alternative approach is the Carbon Maintenance Fee (CMF), which would provide each country with the incentive to encourage the use of biochar along with other aspects of sustainable land management. The CMF would provide the same—or greater—incentive. It is simple, it avoids most of the complications surrounding verification, and it could be applied immediately. But before describing it I will comment on the history of climate negotiations to indicate why an alternative approach is necessary.

Scientists first raised the alarm about global warming in the 1980s. As I said in the Introduction, this led surprisingly quickly to the most remarkable meeting of leaders that has ever taken place. At the Rio Earth Summit of 1992, heads of state from 108 countries with representatives from 64 others got together specifically to listen to scientists. The resulting Framework Convention on Climate Change (UNFCCC) set an agenda, which included the requirement that greenhouse gases must be stabilized "at a level that would prevent dangerous human-induced interference with the climate system." It also said, and this is usually forgotten, that this should be achieved "on the basis of equity."

Cap-and-trade

The Kyoto Protocol was added to the UNFCCC in 1997. It introduced extremely complicated market mechanisms based on carbon trading. Also, the world was divided into two camps: rich participating countries ("annex 1"), and others. The "others" included China and India which have the fastest growth in emissions.

Now, seventeen years after Rio, greenhouse gas emissions have not reduced. They have not even stabilized. Until the credit crunch intervened they were increasing faster than ever before.

Negotiators, influenced by the prevailing belief in the market, introduced commercial incentives for the reduction of emissions through new opportunities for profit: carbon trading. Technically this approach has problems since it requires assessing what comes out of a billion chimneys and exhaust pipes as well as out of the stomachs of cows. They lumped everything together—vehicle exhausts, power generation, soil carbon, biomass, nitrous oxide, methane, municipal waste, forestation, deforestation—into one grand scheme, ignoring the fact that measures required to reduce emissions from burning fossil fuels are totally different from measures relating to land management. This created a bureaucrat's utopia for extended negotiations.

Incorporating biochar into this process would add even more complicated negotiations on what constitutes getting carbon into the ground, the effects of using different kinds of feedstock, emissions related to different techniques for making biochar, how long it would remain in the soil, and a host of other uncertainties. Biochar trading would also require an army of monitors in all countries throughout the world to measure what millions of smallholder farmers might have put into the ground but cannot be seen. Then, would the money get into the right hands? Competing claims would be ubiquitous.

The original intention at Kyoto was for participating countries to achieve emission cuts within their borders. Rich countries feared that actual reductions would harm their industries and make them uncompetitive, so flexible mechanisms were introduced with the intention of achieving the cuts at least cost and least disruption to these industries. "Joint Implementation" (JI) allowed them to trade their emission allowances with other participating rich countries. "Clean Development Mechanisms" (CDMs) allowed participating countries to achieve part of their "reductions" in developing countries that were not subject to the restraints. They would pay for a low-carbon project, say in India, and receive carbon credits that allowed them to emit more than otherwise under their allocation at home. Business could carry on as usual. In theory it does not matter where emissions are reduced, provided they are reduced. If they are reduced in poor countries then the poor get income and the rich can continue to make a lot of money by keeping the global economy growing. Everyone is happy!

Participating nations committed themselves to reduce their emission by 5.2 percent below 1990 levels by 2008. They failed to honor their commitment, and many of the claimed reductions in poor countries did not take place.

Promoters of carbon trading focus on the financial incentive it gives for introducing low-carbon technology. But where does the money come from to pay for these credits? It comes from corporations in participating countries that can increase their wealth by continuing to dominate the industrial field. As they become richer they are able to buy more credits from poor countries. This, in turn, enables their industry to become yet richer and pay more for yet more emission permits. It is a circular process that has allowed emissions to increase. It plays into the hands of the corporations that put so much effort into lobbying the Kyoto negotiators.

The financial collapse caused a slowdown in business, so the cost of carbon credits also collapsed. Corporations have been able to buy them at a knockdown price, and the trade has ceased to be a limiting factor on emissions.

The future of these carbon-trading mechanisms is set to be even more turbulent because of prospects for the global economy. Whether or not JIs and CDMs are extended into the future, it would be a waste of time, and probably ineffective, to try to incorporate biochar.

The global economy

Market mechanisms link non-negotiable environmental imperatives with a faulty human construct: the global economy.

As we have seen recently, the global economy has lost any pretence at stability. One month hundreds of billions of dollars do not exist. The next month, they do exist. The monetary value of assets, like buildings and businesses, has dropped by over $40 trillion around the world. Suddenly the UK has £50,000 ($80,000) debt for every single citizen. The price of oil, on which the modern economy depends, has behaved like a yo-yo. It shot up to $150 a barrel in 2008. Oil-exporting countries had nowhere to invest their trillions so they put them into dead storage. With this money extracted from what Adam Smith defined as the "great wheel of circulation," the global economy collapsed. The price of oil dropped to $34 in 2009. It is rising again and may well exceed the previous high before causing another collapse.

We have a debt-based money system in which interest is charged on all debt. Interest, by definition, means that the economy has to grow, resulting in accelerated depletion of essential but finite resources. The resource and biodiversity crisis is therefore driven from the heart of the Western economy.

It is scarcely believable, but governments allow private commercial institutions (banks) to create almost all the money in use—bank account money—by issuing mortgages to householders and loans to business. Even the government borrows money from banks—and pays them interest—for public works (surely it should be a criminal act for anyone other than the state to create the state's money?). Then, when banks misuse this amazing privilege, the government (taxpayer) bails them out and we are left with fewer and bigger monsters that are too powerful to be allowed to fail.

Banks run a global casino with the mountains of interest they receive, betting on the rise and fall of currencies as well as the stock market. Not content with the interest flowing into their coffers, they call the debts "financial assets" and sell them to unsuspecting investors, thus divesting themselves of liability for bad debts and accumulating yet more money. Then they go into the stratosphere with hedge funds, derivatives, and collateral debt obligations. Ben Bernanke, chair of the Federal Reserve, said before the crash that he didn't understand how hedge funds worked but he had confidence in "sophisticated financial institutions" that did; no wonder the rest of us are baffled. The amount of bank account money is enormous. No one knows how much, but financial commentators reckon it to be at least three times the "value" of real assets in the world. It is known, however, that there is 20 times as much trade in debt (financial assets) as trade in goods and services. Even if the political elite now succeeds in its fantastical attempt to get the economy "back on track" it will be toppled again when the production of oil seriously starts to decline and its price goes up, probably from 2012. The future is deflation, with the value of assets and bank-account money declining.

In his book *The Ascent of Money* Niall Ferguson shows in anecdote after anecdote how, after each collapse, Western banks strengthened their power through their control of money. In the autumn of 2008 Western governments had the opportunity to reform and regulate the financial sector but, due to incompetence or cowardice, failed to do so. The banks are emerging fewer, bigger, and yet again dedicated to running their destructive casino. In his Afterword, Ferguson refers to the ascent as equivalent to Darwinian evolution. But it is evolution by one species for its own benefit. A general law of nature holds that species that are too successful cause their own collapse. The next collapse of Western banking will be even more damaging than the present one. The legacy of debt and banking opulence will be pervasive social unrest.

Put all these things together and the global economy—as developed by Western banks—resembles an asylum run by its inmates.

In his book *When China Rules the World* Martin Jacques argues that Confucian China was always more concerned to avoid social unrest than achieving expansion or commercial prosperity. It used to be thought that China's recent success was simply based on producing goods cheaply for the Western market, and that without this outlet it would fail. It is now decoupling its economy from the West. Asian countries, including China, Russia, and India, have already formed a pact to use their own currencies for reserves and trade. China is asking for an international currency independent of any national currency. John Maynard Keynes proposed this in 1944 but was overruled by the US, which insisted on the petrodollar that made it so wealthy. If China is overruled, it is quite possible that the renminbi will take over from the dollar as the main reserve currency. China may well force a rejection of the Western financial system in the near future.

Proponents of carbon trading expect poor countries to sell emission permits to the West, to the US in particular. But the US is in deep debt. It can only run its industry and wars with reserves deposited by countries like China, Japan, and Saudi Arabia. There is therefore a big question mark over the future of this trade.

With uncertainties hanging over the global economy it would be unwise to make the sequestration of carbon dioxide using biochar dependent on a trade embedded in this financial mayhem.

Experience

But let us consider the actual performance of cap-and-trade. In practice many CDM and JI schemes have been counterproductive. Energy-saving projects planned by the governments of poor countries have been put on hold in the hope that they might earn credits in due course. Other projects that went ahead were not properly maintained or monitored. Some were carried out but subsequently abandoned. So the rich ("participating countries") could increase emissions above their caps while emissions were not reduced elsewhere. Some of these measures, including voluntary offset schemes, involved planting trees in other people's countries: land in poor countries became unavailable to local people because it was set aside to absorb the pollution of the rich.

The list of official and unofficial exploitation of carbon trading goes on. This is not conjecture: I have found that the issue comes up in conversation in India, and many such schemes are reported in the media. Take one instance that has been widely quoted. An offset company visited a township in South Africa and distributed low-energy light bulbs, then left. If a bulb broke the householder could not afford to replace it, nor was the company around to provide the replacement. What's more, the electricity provider had been intending to distribute low-energy light bulbs and their replacements in order to avoid the need to generate more electricity, but this was put on hold when the generous offset offer was made. Poor countries are happy with CDMs and offsets because they bring income regardless of whether they reduce emissions. But the main result and, one suspects, the main intention behind these measures was to maintain the dominance of industry in rich nations. In short, a form of economic imperialism.

The UNFCCC requirement for equity—which should have led to equal rights and benefits for poor as well as rich nations—was totally ignored. At the 1997 meeting and subsequently, Aubrey Meyer pressed again and again for permits to be issued on the basis of one person, one permit, since all people in the world had an equal right to use the properties of the atmosphere. His proposal was simply an extension of other rights such as equality before the law and democracy (one person, one vote). Meyer developed a detailed analysis of how the policy could be realistically applied, and called it Contraction and Convergence. But he was sidelined. The protocol was based on commercial considerations. Period.

The Kyoto Protocol, with its market-based mechanisms, has failed. Since 1997 the annual amount of emissions has been rising at an ever-faster rate and the concentration of greenhouse gases in the atmosphere has increased to dangerous, maybe catastrophic, levels. Between 1970 and 2000 the concentration of greenhouse gases rose by 1.5ppm (parts per million) a year. Since then it has risen at 2.1ppm a year. In 2007 the rise was 2.2ppm. Various excuses have been given for its failure: the US refused to ratify it and Russia withdrew. China and India were not subject to a cap. This does not alter the fact that the agenda set by heads of state in 1992 has not been, and shows no signs of being, implemented.

Market-based economic orthodoxy, therefore, is responsible not only for throwing the global economy into crisis, but also for failing to address the climate crisis.

Smallholder farmers in poor countries manage much of the world's productive land, and it is these that need to be encouraged to produce and use biochar. They have little contact or understanding of the international economy, and many are largely outside any market economy altogether. It does not take much imagination to appreciate that the potential for abuse would be enormous if financial benefits were supposed to be available to these smallholders from international finance.

Europe

If a global system of carbon trading has insurmountable problems due to the vagaries of the global economy and the shift of global power, might a regional one be a useful tool? Many states in the US are introducing these schemes and, being within a single currency and a single administration, they are likely to be effective. But can they be scaled up? Negotiators are looking at an intermediate one—the European Union Emissions Trading Scheme (EU ETS)—for the answer. This spans different economies and administrations.

The ETS was established in 2005 as the world's first proper carbon market. It is limited to large polluting companies that are responsible for less than half Europe's emissions. A cap was put on the total emissions from these companies. Permits, within the cap, were given (not auctioned) to the companies. Companies that want extra permits can buy them from companies that do not use their full allocation. The trade in these permits sets the price for emitting a ton of carbon dioxide.

The gift of permits increases company profits and costs are passed on to customers. Some companies have made huge profits from the trade: British power companies, for example, are making £800 million ($1.2 billion) annually. And it is not effective: carbon emissions in Europe have been rising not falling. Proponents said that this is due to teething problems: too many permits were issued and this would be corrected in the second phase.

The second phase of the ETS is failing for other reasons: those to do with the economy. With the economic downturn, heavy industries mothballed their factories, demand for energy dropped and the price of permits to emit carbon dioxide also dropped. Industries then tried to raise cash by selling their unused permits, flooding the market and depressing prices even further. Previously the price for a permit to emit a metric ton of carbon dioxide was above €30. With the recession it dropped to below €8. This has removed any incentive to de-carbonize the economy. And, in spite of the objective being to wean us off fossil fuels, all kinds of green energy schemes have ground to a halt.

The EU ETS is not a good advertisement for cap-and-trade.

Biochar

Biochar credits could also be self-defeating. If global carbon credits were introduced for burying biochar, if this could be achieved in a reasonable timescale (which is unlikely), and if the practice spread as hoped, the carbon market would be flooded with credits and their price would tumble. Established industries would scoop up the cheap credits and continue with business as usual. These corporations would be in an ever-stronger position to dominate global trade, become increasingly wealthy, and be able to buy up yet more credits. Inequality would intensify. The only way to avoid this would be a hasty reduction in the "cap," but global agreements cannot be modified in haste, particularly if it means setting tougher targets.

Rich countries want credits from biochar to be brought into the Kyoto process because they could be achieved at little cost: actually, farmers would make an operating profit. Three-quarters of the offset potential is in developing countries, so these countries would receive an income. This sets the stage for a cosy agreement between rich and poor for the future of the Kyoto process—the rich continue with business-as-usual and the poor get income—that totally fails to address the needs of the climate.

Money changes hands with carbon credits. This means that permanent sequestration of carbon dioxide must be quantified and verified. Different techniques would have to be assessed. Biochar dust that is spread on the surface might oxidize if not covered. Fields with added biochar may be plowed at some stage in the future and the soil carbon exposed to the air. It has even been suggested that the introduction of biochar may reduce existing soil carbon in some circumstances. Biochar that is made from trees releases one form of locked carbon and transfers it to another form, and it would be difficult to measure the net benefit. The rising value of credits would lead to monoculture land management and the destruction of existing diverse woodlands. The overall effect of this on carbon sequestration might be negative, but would certainly be difficult to assess; it would also reduce biodiversity. These are just a few of the uncertainties that would have to be determined in regulations for claiming carbon credits. Failure to define what constitutes sequestration accurately would lead to perverse commercial outcomes. Failure to achieve verifiable rules would allow the trade to be contested. These issues indicate that negotiations to establish a global regime of carbon credits would take years and, in effect, be dependent on an international police force.

Local cap-and-trade schemes within a currency and within a single administration might be viable in order to help a state achieve its own carbon goal. This would, however, not reduce net emissions if offset credits are used simply to bring the state's emissions down to the level of its global allocation, since what is achieved with biochar would be lost with increased industrial emissions.

Twin solutions

If biochar is to be regulated for the sequestration of carbon dioxide through being an integral part of farming, then the incentive must be found for governments to take direct action themselves. This is the purpose of the Carbon Maintenance Fee that I describe below.

The human-induced increase in the atmospheric concentration of greenhouse gases has two main sources. Two-thirds are currently coming from burning fossil fuels, one-third from the management of land. The attempt to regulate the two within a single framework gives rise to endless and unnecessary complications. If kept separate, regulations can be simplified and made more effective. Biochar falls into the second category.

The simplest way to control emissions from burning fossil fuels—the first category—is to put a cap on the amount of coal, gas, and oil mined. It would not then be necessary to attempt to measure emissions because the amount of these fuels brought onto the market would determine the amount of carbon dioxide eventually released. This "upstream" regulation would mean that development would take place using the fuel that is available, regardless of the state of the global economy or the needs of different countries. The requirement for equity should also be met upstream: everyone in the world would be given an equal entitlement (adding up to the cap), and mining companies would buy these through brokers before selling their fuel. They would not be allowed to sell more than covered by the entitlements they acquire. The huge sums that energy companies receive would then be distributed at grass-roots level to keep a stable, more equitable, global economy. There are only about 500 extraction companies, and it would be easier to regulate these than monitor billions of chimneys and exhausts—the present end-of-pipe approach. This "upstream" equitable regulation is called cap-and-share in Europe and cap-and-dividend in the US.

Emissions caused by the way land is used—the second category—including, for example, those from deforestation, flooded rice fields, plowing soil, keeping cattle, and the use of nitrogenous fertilizers, should be subject to a separate regulation. The use of biochar would be covered in this category.

A clause in the Kyoto Protocol (article 3.3) stipulates that emissions from and absorptions into land can only be counted if they can be measured as "verifiable changes in carbon stocks." At the time of Kyoto this could not be done for whole countries because appropriate technology did not exist. Instead, the attempt was made to assess the effects of each human interaction with the land separately. This led to serious problems of definition and verification, and also meant that slow changes to the global environment, perhaps as result of global warming, were ignored. Many countries feel that the attempt to base assessment on individual activities should be abandoned, now that the overall content of carbon in each country can be measured.

CMF: the Carbon Maintenance Fee

Remote sensing by satellite, linked to soil sampling, can measure the amount of carbon in plants and trees (above ground biomass), roots (below ground biomass), cut wood and litter (dead organic matter), and soil carbon. Surveys could be carried out annually at the same time of year and, after they have been calibrated by on-ground sampling, they would measure the increase or decrease in the carbon pool of a country (a carbon pool is defined as "a system with the capacity to accumulate or release carbon"). The satellite survey will provide the extent of each type of ground cover together with its mass, and the sampling will give the typical carbon content of the soils under it and how they are changing over time. Although the size of plots that can be monitored is continually reducing, it is not possible—for some of the reasons given above—for remote sensing to assess individual activities.

The UN Food and Agriculture Organization has prepared a Global Soil Database with much of the necessary information. The most advanced application seems to be in New Zealand, with the Land Use and Carbon Analysis System (LUCAS). This was set up in 1990. The initial survey was carried out five years ago and the first remeasurement is due shortly. This will identify changes in the country's carbon pool.

The Australian Carbon Accreditation Scheme (ASCAS) makes annual incentive payments for the increase of soil carbon above an initial baseline. These payments, together with the ability to compile validated data, are almost identical to the measures proposed for the Carbon Maintenance Fee. Christine Jones, of Amazing Carbon, says that up to 80 percent of carbon has been lost from most farmed soils in Australia since Europeans settled, and she contests the attitude of many scientists that soil can't be restored. "The quickest and most cost effective way to restore degraded cropland is through a grazed perennial pasture ley," she says. This is in line with Graham Harvey's analysis of the American prairies in his book *The Carbon Fields*, where he says that before cultivation the land held more than ten times as much carbon as it does now. Biochar's ability to retain moisture and support microbial activity makes it an obvious tool for helping to restore vast areas of degraded land in both Australia and the USA.

The Carbon Cycle and Sinks Network in Ireland proposes that a global survey should be carried out to provide a baseline against which future measurements can be compared. Since it would be in a country's interest for the baseline to show a low content, it should be put in hand immediately as otherwise initiatives that could increase the carbon pool will be delayed. The initial survey could be carried out in advance of international agreement on monetary arrangements for the CMF. Most of the information already exists. Indeed monetary agreements would be easier to achieve if the data were available. The survey should be repeated at regular intervals in order to monitor the carbon that has been gained or lost.

Each country would then be paid a fee for maintaining the carbon that is contained in its biomass and soil, the "Carbon Maintenance Fee" (CMF). Any country showing an increase would receive an additional payment. Any country showing a loss of carbon would have to pay a price for each ton lost.

Separate methods would be used to assess the emissions of nitrous oxide, and these would be incorporated in the calculation for each country. Other non-carbon greenhouse gases are not sufficiently pervasive to require inclusion.

The Carbon Cycle and Sinks Network would not require special arrangements for the application of biochar and, if its approach is adopted, almost all the uncertainties associated with measuring the net benefit of biochar would be removed. There would, for example, be no need to measure the stability of biochar in the soil or on the surface (differing views have been expressed on whether it would degrade) since this would be part of a gain or loss in a country's carbon pool established by remote sensing and periodic surveys. Buried biochar made from plant material that had captured carbon dioxide while growing would provide an increase in the carbon pool at each review. Trees planted (whether for future biochar or not) would provide a carbon gain. When trees are converted into biochar there would be a carbon loss. Emissions made during the pyrolysis process would also result in a carbon loss. The net gain that is anticipated from the use of biochar would simply add to the country's carbon pool in the same way that sustainable agriculture would add to its carbon pool.

With the Carbon Maintenance Fee it would then be in a country's interest to increase its carbon pool. It would be in its interest to maintain any mature woods and forests, to encourage farming that increases soil carbon, to find and subsidize appropriate equipment for the production of biochar (for example, to subsidize pyrolysis stoves in preference to wood-burning stoves) to set up advice centers that can help small-scale farmers and the informal sector (for example, to find the right balance between using farm waste for compost versus biochar). The more a country can induce its businesses and citizens to increase the carbon pool, the more it will benefit from the Carbon Maintenance Fee.

The Kyoto Protocol tried to make its rewards and penalties self-financing through the trading process. The emissions reduction targets set for the rich countries created a value for carbon, thus providing an incentive to reduce its emission. The same route could be followed for funding the Carbon Maintenance Fee, though it would need to take into account the past failures of this market-based approach. If cap-and-share is adopted for fossil-fuel emissions, a proportion of allocations could be held back and auctioned to fossil-fuel extractor companies to provide the fee, or part of it. However, since global warming is the greatest threat humanity has ever faced, there should be many other ways in which money can be generated to fund the fee, such as a tax on capital movements. The higher the fee, the more incentive each country would have to increase its carbon pool, and the greater the possibility that we might avoid the more catastrophic effects of global warming. The fee should become a major element of global finance. Les Carter, a correspondent from Canada, put it like this: "We don't 'need' more energy, we don't 'need' economic stimulus, we don't even 'need' jobs. What we do need is a stable climate."

Chapter 8

Summary

Charcoal has been with us for millennia. Biochar—finely crushed charcoal used for soil enhancement—is a new enthusiasm.

All plants capture carbon dioxide from the air by photosynthesis, and microorganisms release it as the plants rot. All plant material, not just wood, can be turned into charcoal or biochar (I apologize to scientists for using the imprecise "all"). So putting biochar into the soil extracts a greenhouse gas from the air. This can help to avoid the worst effects of global warming.

The proper management of land, particularly through sustainable systems like organic farming, permaculture, and forest gardening, also allows carbon to be retained in soils. These practices, together with biochar, appear to be the only immediately available and tested means of extracting carbon dioxide from the atmosphere.

Rich dark areas of deep soil in the Amazon rainforest, where surrounding land is thin and infertile, demonstrate the ability of charcoal to improve soils and retain carbon permanently. The civilization that produced these soils was destroyed by a pandemic brought from Europe 500 years ago.

Pilot schemes are now testing ways to create carbon-rich soils using biochar with varying degrees of success. Scientists are working on details of the process. There is now a need for an organization to advise practitioners, particularly farmers, on best practice in the way that organic growers can obtain advice and certification.

Industrial farming will undergo fundamental change due to peak oil and peak phosphorus. In Britain, organic farming can maintain present levels of food production, but increased production will only come from the informal sector. This will require more people growing food:

research and development should therefore concentrate on making the benefits of biochar available to this sector. In countries with large rural populations—in line with the recommendations of the UN Food and Agriculture Organization—smallholders need to be served with research, appropriate equipment and viable markets by their governments so that they are not driven from the land. Biochar will have most effect in tropical areas, where oxidation of soil organic matter is greatest and where much degraded soil and near-desert land exists.

Charcoal dust is light in weight when first produced, because it is riddled with microscopic holes that gradually capture and retain moisture. This provides an immediate benefit, particularly in arid parts of the world.

The introduction of biochar into soil is not like applying fertilizer; it is the beginning of a process. Most of the benefit is achieved through microbes and fungi that colonize it and integrate it into the surrounding soil.

When outcomes are small you can afford to be relaxed about taking risks. Potential outcomes now are not small. Global warming might make life difficult, or it could cause our extermination. Looming food shortages could lead to mass starvation and widespread conflict. Biochar has a major role to play in both fields.

Emissions from burning fossil fuels can best be controlled "upstream" by limiting the amount of these fuels entering the economy: *globally* by controlling the amount of coal, gas, and oil that is extracted from the ground; *nationally* by controlling the amount of the fuels entering the country. The mechanism for achieving this is called cap-and-share in Europe and cap-and-dividend in the US.

Emissions from land-based activities can best be controlled on an annual basis through remote sensing of the carbon contained in plants, soils, and roots. This is a relatively new technology that is carried out by satellite, though New Zealand has been monitoring its land for the last five years. The Irish proposal for its use is called the Carbon Maintenance Fee (CMF): governments are given a fee for maintaining the carbon pool of their land; they are rewarded if the carbon pool increases and penalized if it reduces. The fee should be a major element in world finance. In order to get the benefits of CMF governments will need to find ways to enable all businesses and citizens to sequester carbon. The best, and possibly the only, ways to achieve this are through incorporation of biochar into soil together with sustainable farming. The CMF can and should be started without delay.

The suggestion that carbon credits should be applied to biochar gives rise to many problems, particularly over verification. The net effectiveness and permanence of biochar in the soil are subject to a whole host of variables that would lead to dispute and delay. I have also outlined unforeseen effects that might result. The CMF would provide the same incentives while avoiding the pitfalls.

Some readers will be asking: "What can I do?"

If you have a garden, allotment, or smallholding, you have land into which you can incorporate biochar. Many individuals are experimenting with equipment. An oil drum on its side is sometimes used (you can find examples on the internet). Some have obtained the Anila stove. Maybe other techniques will emerge. My fear is that this will be a temporary enthusiasm until the tedium of managing it kicks in. For more lazy people like me it would be better if biochar, already charged with compost, were readily available at garden centers, or if municipalities convert their urban waste and distribute it. An enlightened government that wishes to increase its carbon pool would provide large incentives for encouraging this practice, and under CMF the government would be reimbursed many times over. If all gardens, all allotments, all smallholdings were getting biochar dug into their land the national carbon pool would increase fast. This process will be speeded up when large farms are broken down into smaller units—due to peak oil—and a much larger proportion of the population turns its hands to cultivation.

Due to economic turmoil—another result of peak oil—unemployment will rise. If you lose your job, returning to the same employment may not be an option: mainstream employment is on a downward path. Instead, retrain in organic cultivation. Your country needs you!

James Lovelock spoke only about farmers; he should have included the whole of the informal sector, you and me, when referring to the dangers of global heating: "There is one way we could save ourselves and that is through the massive burial of charcoal. It would mean farmers turning all their agricultural waste—which contains carbon that plants have spent the summer sequestering—into non-biodegradable charcoal and burying it in the soil. . . . This scheme would need no subsidy: the farmer would make a profit."

Selected Bibliography

General

Ed Ayres, *God's Last Offer: Negotiations for a Sustainable Future*, Four Walls Eight Windows, 1999.

James Bruges, *The Big Earth Book: Ideas and Solutions for a Planet in Crisis*, 2nd ed., Sawday, 2008.

Richard Douthwaite, *Short Circuit: Strengthening Local Economies for Security in an Unstable World*, Green Books, 1996.

Fiona Harvey, "Black is the New Green," *Financial Times*, 2009.

Rob Hopkins, *The Transition Handbook: From Oil Dependency to Local Resilience*, Chelsea Green Publishing, 2008.

Martin Jacques, *When China Rules the World: The Rise of the Middle Kingdom and the End of the Western World*, Allen Lane, 2009.

Johannes Lehmann and Stephen Joseph (editors), *Biochar for Environmental Management: Science and Technology*, Earthscan, 2009.

Mark Leonard, *What Does China Think?* Fourth Estate, 2008.

James Lovelock, *Gaia: A New Look at Life on Earth*, Oxford University Press, 2000.

James Lovelock, *The Vanishing Face of Gaia: A Final Warning*, Basic Books, 2009.

Mark Lynas, *High Tide: The Truth About Our Climate Crisis*, Picador, 2004.

Daniel Quinn, *Ishmael: An Adventure of the Mind and Spirit*, Bantam/Turner, 1992.

E. F. Schumacher, *Small Is Beautiful: A Study of Economics As If People Mattered*, Blond & Briggs, 1973.

Joseph A. Tainter, *The Collapse of Complex Societies*, Cambridge, 1988.

Agriculture

Alison Benjamin and Brian McCallum, *A World without Bees*, Guardian Books, 2008.

Rachel Carson, *Silent Spring*, Houghton Mifflin, 1962.

The Community Solution, *The Power of Community: How Cuba Survived the Oil Peak*, DVD, 2008.

Corporate Europe Observatory, *Reclaiming Public Water: Achievements, Struggles and Visions from Around the World*, Transnational Institute & CEO, 2005.

Grace Gershuny and Joe Smillie, *The Soul of Soil: A Soil-Building Guide for Master Gardeners and Farmers*, 4th edition, Chelsea Green, 1999.

Graham Harvey, *The Carbon Fields*, GrassRoots, 2008.

Rebecca Hosking, *A Farm for the Future*, BBC Natural World, 2009.

Sir Albert Howard, *An Agricultural Testament*, Other India Press, 1940.

Christine Jones, "Australian Soil Carbon Accreditation Scheme," www. amazingcarbon.com, 2007.

Philip Jones & Richard Crane, *England and Wales Under Organic Agriculture: How Much Food Can Be Produced?* Centre for Agricultural Strategy Report 18, University of Reading, 2009.

Felicity Lawrence, *Not on the Label: What Really Goes into the Food on Your Plate*, Penguin, 2004.

E. W. Russell, *Soil Condition and Plant Growth*, Longman, 1912 and subsequent editions.

Charlie Ryrie, *Soil*, HDRA & Soil Association, Gaia Books, 2001.

Peter Singer, *In Defence of Animals*, Blackwell, 2006.

Colin Tudge, *So Shall We Reap*, Allen Lane, 2003.

Edward O. Wilson, *The Diversity of Life*, Harvard, 1992.

Energy

Kenneth S. Deffeyes, *Hubbert's Peak: The Impending World Oil Shortage*, Princeton, 2001.

Richard Douthwaite (editor), *Before the Wells Run Dry: Ireland's Transition to Renewable Energy*, Feasta, 2003.

Richard Heinberg, *Powerdown: Options and Actions for a Post-Carbon World*, New Society Publishers, 2004.

David J. C. MacKay, *Sustainable Energy—Without the Hot Air*, UIT, 2009.

Hermann Scheer, *Energy Autonomy*, Earthscan, 2007.

Pat Thomas (editor), "Biofuel: Special Report," *The Ecologist*, March 2007.

Economy

Richard Douthwaite, *The Ecology of Money*, Schumacher Briefing No. 4, 1999.

Richard Douthwaite, *The Growth Illusion: How Economic Growth Has Enriched the Few, Impoverished the Many and Endangered the Planet*, Green Books, 1992.

Niall Ferguson, *The Ascent of Money: A Financial History of the World*, Penguin, 2009.

Hazel Henderson, *Beyond Globalization: Shaping a Sustainable Global Economy*, Kumarian Press, 1999.

Joseph Huber & James Robertson, *Creating New Money: A Monetary Reform for the Information Age,* New Economics Foundation, 2000.

The Liquidity Network, Feasta (in preparation) 2009.

Michael Rowbotham, *Goodbye America! Globalisation, Debt and the Dollar Empire,* Jon Carpenter Publishing, 2000.

Michael Rowbotham, *The Grip of Death: A Study of Modern Money, Debt Slavery and Destructive Economics,* Jon Carpenter Publishing, 1989.

Joseph E. Stiglitz, *Globalization and Its Discontents,* W. W. Norton, 2002.

Regulation

AEA, *Environment and Energy: Cap & Share,* AEA report, 2008.

Peter Barnes, *Climate Solutions: What Works, What Doesn't and Why,* Chelsea Green Publishing, 2008.

Carbon Maintenance Fee, Feasta (in preparation), 2009.

Richard Douthwaite, *Revising the Programme for Government* (in preparation), 2009.

Richard Douthwaite and Corinna Byrne (editor), *Reducing Greenhouse Emissions from Activities on the Land,* Feasta for CCSN, 2009.

Larry Elliott and Dan Atkinson, *The Gods That Failed: How Blind Faith in Markets Has Cost Us Our Future,* The Bodley Head, 2008.

Global Soil Database: Including Potential to Sequester Additional Carbon in Soils, UN FAO, 2008.

Mayer Hillman, *How We Can Save the Planet,* Penguin, 2004.

Jeremy Leggett, *The Carbon War: Dispatches from the End of the Oil Century,* Allen Lane, 1999.

LUCAS: Land Use and Carbon Analysis System, Ministry for the Environment, New Zealand.

Aubrey Meyer, *Contraction & Convergence,* Schumacher Briefing No.5, 2000.

Sara I. Scherr & Sajal Sthapit, *Mitigating Climate Change Through Food and Land Use,* Worldwatch Report 179, 2009.

SEDAC: Socioeconomic Data and Applications center, including *Human Appropriation of Net Primary Productivity (HANPP),* http://sedac.ciesin .columbia.edu/es/hanpp.html and http://sedac.ciesin.columbia.edu.

Amartya Sen, *The Idea of Justice,* Harvard, 2009.

Peter Singer, *One World: The Ethics of Globalization,* Yale, 2008.

Gerard Traufetter, "Planting Trees to Atone for Our Environmental Sins," *Der Spiegel,* 2006.

Index

About the Author

James Bruges worked as an architect in London, Sudan, and India until 1995 when he retired in order to write about economic and environmental issues. He is the author of *Sustainability and the Bristol Urban Village Initiative, The Little Earth Book*, and *The Big Earth Book*, and was a contributor to *What About China?* His work has also appeared in *Resurgence, The Friend*, and *The Ecologist*. He was raised in Kashmir until the age of twelve and now lives with his wife, Marion, in Bristol, England.

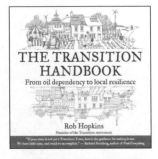
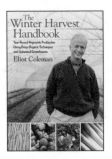
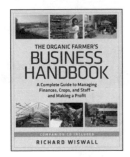